New Technologies for Energy Efficiency

New Technologies for Energy Efficiency

by
Michael F. Hordeski

THE FAIRMONT PRESS, INC.
Lilburn, Georgia

MARCEL DEKKER, INC.
New York and Basel

Library of Congress Cataloging-in-Publication Data

Hordeski, Michael F.
    New technologies for energy efficiency/by Michael F. Hordeski
        p. cm.
    Includes bibliographical references and index.
    ISBN 0-88173-369-5 (print)
        1. Energy conservation--Technological innovations. I. Title

    TJ163.3 .H67 2002
    621.042--dc21   2002029849

Published by The Fairmont Press, Inc.
700 Indian Trail, Lilburn, GA 30047
tel: 770-925-9388; fax: 770-381-9865
http://fairmontpress.com

Distributed by Marcel Dekker, Inc.
270 Madison Avenue, New York, NY 10016
tel: 212-696-9000; fax: 212-685-4540
http://www.dekker.com

Printed in the United States of America

10 9 8 7 6 5 4 3 2 1

0-88173-369-5 (The Fairmont Press, Inc.)
0-8247-0936-5 (Marcel Dekker, Inc.)

Dedicated to Paul Schweich

"Keep on flying, Paul."

TABLE OF CONTENTS

PREFACE

The electricity crisis is real in California and several Western and Eastern states today, and may threaten the rest of the nation in the future. The needs of the new economy are stressing an outdated energy infrastructure and affecting the level of electrical service.

The current regulatory system encourages large powerplants with long lead times. The right response to the crisis is to invest in new technologies that reduce the need to depend on the power grid.

There are tools and technologies available for an energy revolution that can imitate the evolution that has occurred in the computer and telecommunications industries.

Fuel cells use the chemical reaction from a variety of fuels to create power and allow companies to generate clean high-grade electricity on site without air pollution problems. Natural gas distributed generation uses gas turbines and gas engines. It places small modular power units close to electric power users. Other distributed generation sources such as solar-voltaic and micro-hydro are also of interest.

Cogeneration systems are also available to small scale users of electricity. These modular systems produce electricity and hot water from engine waste heat. Home sized cogeneration packages are capable of providing most of the heating and electrical needs of a home. Cogeneration can produce a given amount of electric power and process heat with 30% less fuel than it takes to produce the electricity and process heat separately.

These new technologies are equivalent to the use of wireless cell phones and portable computers that are replacing traditional wire-connected phones and desktop computers.

The electrical grid dates back to the beginning of the last century and is out of sync with the new information technologies. The architecture of the existing electrical transmission grid is in opposition to the concept of distributed networks made possible by the Internet.

The best way to insure electrical supply reliability and reduce long-term costs is to utilize these smaller, clean, more efficient energy-generating technologies into your energy management plan.

This book examines the role of new technologies in reducing operating costs and developing more innovative and practical approaches to energy management. It includes developing alternative energy programs, monitoring rates, quality and energy policies, cost effective power generation solutions, small modular power generation units, cost effective energy services, advanced technologies and products, information monitoring and diagnostic systems, air monitoring, energy storage options and lighting and cooling integration.

Chapter 1 is concerned with energy demand, sources and rate trends. It considers energy usage trends and the power future. Basic concepts of energy, power, conversion and efficiency are discussed. Energy pricing and rate structure philosophy are defined. Energy management programs, energy coordinators and energy audits are introduced as solutions.

Chapter 2 discusses energy sources including heat pumps, solar energy and wind power. Heat sources for heat pumps are considered along with passive solar heating, solar collection and solar heat storage. Hot water systems and solar cooling are also discussed.

Chapter 3 considers the integration of cooling, heating and power systems. Topics include energy and power management, distributed control trends and air monitoring. It introduces power quality and lighting upgrade trends along with load shedding and demand side limiting concepts. Cogeneration is also introduced along with district heating and seasonal energy storage.

Chapter 4 explains the underlying concepts of alternating voltage and current. This includes the use of power factors, inductive loads and capacitors. Harmonics are introduced along with the crest factor.

Harmonic Distortion is considered in depth with K-Factor transformers and the h factor. The problem of oversizing transformers is explained. Power system measurements are discussed along with adjustable capacitor banks and harmonics studies using computer simulation. Harmonic filters are considered along with transformer inrush current and capacitor effects. Backup power systems using chemical batteries, kinetic energy storage and superconducting magnetic energy storage are discussed.

Fuel cells will be an important source of power in the future as explained in Chapter 5. Topics include fuel cell technology and characteristics. The problems of different types of fuel cells for electric power production are discussed. Fuel cells may get their greatest boost as replacements for batteries for portable equipment and electric vehicles. Some of these applications may use fuel processors while others would require hydrogen fuel in a hydrogen economy. Fuel cells are compared and their thermodynamic properties discussed.

Chapter 6 considers modular power generation including turbines, gasification and combined cycle generation. Topping and bottoming systems are explained. Turbine controls are discussed along with compressor surge and auxiliary equipment. Biofuels are considered as well as combined cycle technology. Peak reduction is discussed as well as standby rates.

Lighting retrofits are the subject of Chapter 7. Energy savings are possible from simple lamp replacement, upgrading controls and fixtures and repositioning lights. Load management is discussed along with lighting management and lighting audits. Task lighting improves the efficiency of lighting and often results in a reduction of the generated power needed for lighting. Sulfur lamps and light pipes are among the newer techniques used to provide outdoor lighting.

Chapter 8 examines the effect of computer technology on energy control trends. Building management now depends on a network like BACnet or LonWorks. Building management software can automatically implement operation and maintenance schemes for increased efficiency and reliability. The advances in metering and power monitoring are examined as well as safety, security and building air quality.

Many thanks to Dee, who did much in getting both the text and the drawings in their final form.

CHAPTER 1

ENERGY DEMAND, SOURCES AND RATE TRENDS

ENERGY USAGE TRENDS

In California, rolling blackouts have darkened some of the richest cropland, most complex high-tech factories and busiest streets in America. These blackouts were part of a statewide energy crisis that had been years in the making and was triggered by a flawed deregulation plan. It enraged consumers and businessmen alike and pushed California's two largest utilities to insolvency. It was a threat to the state's $1.3 trillion economy which is the sixth largest on Earth and the total U.S. economy as well. The California crisis was part of a nationwide energy problem where natural gas rates have risen to the highest level in years.

San Diego was the first city to encounter the dynamics of an open, competitive power market. It tested California's ability to provide reliable, affordable power in a state whose economy is increasingly dependent upon electricity. Power bills jumped 300% in San Diego for many residential customers. The Cambridge Energy Research Associates stated that the California crisis brings up many questions about our entire energy infrastructure.

The California energy deregulation plan was supposed to cut electric rates, but instead it more than tripled what some California consumers must pay due to a tenfold increase in wholesale prices. Instead of taking regulators out of the utility business, the plan has done just the opposite.

Consumers, power suppliers, legislators and regulators must share some of the blame. Consumers are asked to conserve more since California failed to complete a single large powerplant over a period of 10 years, while California's high tech business boomed and the overall state economy expanded by 34%.

New construction was one of the aims in 1995 when the state launched the nation's first and most sweeping electric deregulation plan which was endorsed by utilities and lawmakers as well as environmental and consumer-advocate groups. The goal was to break up the major monopolies of Pacific Gas & Electric (PG&E) and Southern California Edison (SCE). They would be allowed to purchase and market power in California as well to do business in other states. PG&E now owns 30 plants outside California. In 1998, it paid a then-record $1.8 billion for a plant near Pittsburgh. SCE paid $4.8 billion for plants from Chicago's Commonwealth Edison. SCE gets more than half its earnings from deregulated plants in other states. Its Mission Energy unit owns more than 40 global plants.

Out-of-state operators were expected to bring down electric rates in California which were among the highest in the nation. But, California dismantled its private power-generating industry without securing adequate power supplies. The three largest utilities, which include PG&E, SCE and San Diego Gas & Electric, sold generating plants to outsider utilities like Duke Energy of Charlotte, NC, and Reliant Energy of Houston. PG&E sold more than $1.5 billion of their generating plants and SCE sold about $1.2 billion.

The state would not allow long-term purchasing agreements based on the fear that they would be locked into fixed-price contracts as prices dropped. Power purchases had to be made on the spot or cash market where prices were low at the time.

The utilities accepted this limitation, along with a rate freeze. For a while, they enjoyed making a profit, but the state's demand for electricity was growing fast and its generating capacity was not growing bigger. California imported almost 25% of its power from suppliers in the Southwest and Pacific Northwest. This connection has been torn with the growing demand in those areas. There was also a price spike in natural gas which is used in about half of California's generating plants and accounts for more than half of the price of electricity. As the rising demand for power met its restricted supply, the wholesale price of energy jumped from less than 5 cents per kWh in January 2000 to nearly 40 cents per kWh in December 2000.

The results were predictable. Unable to pass along high costs to homes and business, PG&E and SCE sustained losses and owed more than $12 billion to their banks and power providers. Generating companies became reluctant to supply them with more power.

California's power demand has grown nearly 25% since 1995. Texas has built 22 new plants since 1995, with 15 more coming on-line during 2001. California's Independent System Operator (ISO) manages the power grid and must find about 6,000 megawatts a day outside the state.

Typical of the frustrations that power companies faced was the proposed Coyote Valley generator plant that Calpine wanted to build in San Jose, in the heart of the Silicon Valley. The plant would light 600,000 homes in a region where the City Council vetoed the project.

The plant also faced opposition from Cisco Systems, a leading producer of Internet hardware, which was also San Jose's largest employer. Cisco claimed that the powerplant would be an eyesore next to an industrial park that the company planned to build for some 20,000 employees. Over 40 plants which would add 22,600 megawatts of generating capacity are proposed for California.

About 40% of the state's capacity comes from facilities built more than 30 years ago, making them prone to equipment problems. Compounding the problems of California was a shortage of rain and snow in the Pacific Northwest, which depleted hydroelectric plants of the water they need to generate exportable power. The California grid lost about 3,000 megawatts.

Northern California was also squeezed by bottlenecks in supply lines. Path 15 runs through the state's Central Valley and places a constraint on the northern half of the state that makes blackouts more certain under high demand conditions.

The power grid's hourly needs are projected by computers. Weather is a major factor since a few degrees can make a difference in power requirements. Supply and demand can shift rapidly and companies may have just moments to complete a process or project before they must cut their power. The crisis heightens when workers return to their houses and turn on lights, computers, TVs and cook dinner while the clothes dryer is on. If they waited until 8:00 p.m. to use some of those appliances it would have a major impact.

When shortages persist for more than an hour, the utilities darken another area. At the San Francisco Zoo, the staff had to rush an endangered species of 150 fish from Madagascar to the zoo hospital to rescue them from cold water.

Many states are opting for the merits of a free market for power but will they be plagued by blackouts, near bankrupt utilities and high electricity bills? Nearly half of the states including New York, Pennsylvania, Virginia, Texas, Arizona and Oregon are in various stages of deregulating their power delivery system.

New York City faces an electricity shortage and in the Northwest, Oregon and Washington, which have imported power from California in the winter, have found themselves exposed to a limited supply. With demand booming, reserves have declined from 40% excess two decades ago to 15% today.

The nation's old outdated transmission grid is not built to handle so much long-distance traffic. Increasing demand does not necessarily attract new suppliers since the cost of entry, multibillion-dollar powerplants and years of waiting for approval and returns on investment is not attractive. This forces real competition away. A few states, such as West Virginia, Oklahoma, Arkansas, Nevada and New Mexico, have already halted the process of deregulating.

The Center for the Advancement of Energy Markets argues that the California downfall is not an indictment of electricity deregulation but a lesson on how not to do it. Pennsylvania started deregulation about the same time as California. It is often cited as a model of deregulation. More than 500,000 customers, 10% of the state's total, have switched to one of the new suppliers and the state has seen a $3 billion savings on electric bills. Pennsylvania utilities were not required to sell the bulk of their powerplants and become middlemen. They were also allowed to lock in long-term contracts instead of relying on the daily spot market.

Pennsylvania also ensured that there were enough players to make a truly competitive market (200 buyers and sellers). Instead of trying to go it largely alone as California did, Pennsylvania is part of an integrated, five-state trading market, including New Jersey and Maryland.

Texas operates its own electricity grid, which makes it less vulnerable to bottlenecks in the national system. It also imports less than 1%

of its power. Like Pennsylvania, Texas did not require utilities to sell off their plants, and it did allow long-term contracts. Texas has encouraged powerplant construction. Its environmental regulations are less strict than California's, and its approval process is more streamlined.

Since 1995, over 20 new plants have come on-line and almost that many should be up and running in a few years. With that much capacity, state officials have guaranteed a 6% rate cut with retail deregulation taking effect.

Other deregulated states have experienced price movements in the other direction. In Massachusetts, consumers were promised 15% cuts, but rates have increased as much as 50%. A report from the Union of Concerned Scientists inferred that this might not be due entirely to the result of market forces. Since deregulation began, plants have been shut down for maintenance nearly 50% more often than before. Producers point to stricter environmental rules as the reason for the increases.

New York City and parts of nearby Westchester County were among the first areas in the country to have retail rates entirely deregulated. Rate increases have been as much as 30%. Much of the increase is due to rising costs of the gas and oil that power generators in the Northeast use. This forced the New York State public service commission to consider the implementation of a temporary, $150 per megawatt-hour wholesale price cap.

In New York City, old transmission lines are strained to bring in enough power to foster the booming economy. Getting a power plant built anywhere near the city can be a difficult task. In the last 15 years before the state deregulated, only one new large plant went up. There are now dozens of applications in the pipeline. In spite of community opposition, the New York Power Authority has placed ten 44-megawatt natural-gas-powered generators around the city.

New York's approach is just one of many that states are using. Ohio opened up its market in 2001 and the state is encouraging groups of customers including schools, churches, neighborhoods and cities to band together in pools to negotiate with suppliers. In New Jersey, one of the new energy providers closed because of high costs, so officials are trying to reduce the paperwork required for a customer to leave their old utility. Some New England states require utilities to keep adequate reserves to limit supply problems.

More innovative measures may be needed to allow deregulation to work. Consumers might choose from a set of customized electricity packages, in much the same way as all the long-distance offerings. With the help of the Internet as a real-time monitoring tool, consumers could use their washer or dryer at the cheapest time of the day. The analog meters at people's homes are an archaic part of a century old power system. Suppose we had free Sundays or late night rates?

The North American Electric Reliability Council estimates the U.S. will have enough electricity to keep up with demand even if only half of all the proposed plants are actually built. But, many may not make it through the approval process.

A national energy consulting firm, Cambridge Energy Research Associates, describes the power industry in the United States as being in a radical shift from regulatory to competitive market forces. In California, the shift may be in reverse, away from markets and back to government intervention. At a time when power was never more critical to the state's and nation's economic growth, a reverse restructuring is occurring in California where economic theory and political reality are at odds. The state's powerful investor-owned utilities are trying to achieve a balance between pro-market and pro-government forces.

The seeds of California's energy problems were planted long before deregulation. Environmental concerns, poor planning, and uncertainty about the rules of deregulation forced California's power supply to grow far slower than demand, which was driven by an expanding population and its increasing use of energy using technology.

In 1996, a Republican governor and Democratic-controlled state Legislature came together to push electricity deregulation in AB 1890, a bill that went through both houses without a dissenting vote in less than a week's time. The flawed 1996 deregulation scheme exacerbated the problems. It included a pricing system that prevented utilities from locking in long-term supplies at fixed prices and from passing on higher fuel costs to customers. Without rising prices, consumers had no incentive to reduce demand and utilities could not pay their soaring wholesale power bills. That scared away some power suppliers and created the financial crisis.

The biggest flaw in California's deregulation plan was the decision to force utilities to buy all of their power needs one day in advance from a newly formed entity called the California Power Exchange. It was hoped that the exchange would provide the most transparent prices, since every buyer and seller had to operate through it. Any power that did not get bought through the exchange would be purchased on a last-minute basis the following day by another entity called the Independent System Operator (ISO).

This two-step arrangement encouraged generators to offer less power to the exchange and instead to wait and sell power to the ISO, which would have to pay higher prices. Lawsuits by consumer groups and others allege that the electricity marketers engaged in unlawful market manipulation but investigations have failed to prove collusions. One investigation concluded that the soaring prices could not be explained by such factors as high demand and rising fuel and environmental costs. It was concluded that power was being withheld inexplicably, at the exact time at which prices were most vulnerable to manipulation.

Long-term contracts would take the uncertainty out of pricing and the Power Exchange is being phased out. One of the hardest-hit power users in California was California Steel Industries Inc., a steel producer that lost power 14 times.

The path to full competitive markets could take a detour in which the government steps in until the path eventually rejoins a smoother, improved road to deregulation. Congress is unlikely to provide a national electricity deregulation bill, the push will continue to be toward more market influence and less regulatory intervention.

In California the main players in deregulation have been the governor and several state agencies, California Public Utilities Commission, California Energy Commission, California Independent System Operator, California Power Exchange and the California Electricity Oversight Board. Private-sector power generators include AES Corporation, Duke Energy, Dynegy Corporation, NRG, Reliant Energy, Southern Company and the Williams Company.

THE POWER FUTURE

Some economists see the rest of the West using well-functioning wholesale markets with California off on its own. California business officials believe in a free-market power structure, but support government intervention to stabilize wholesale prices and curb market power.

In the short term there is too much market power that needs curbing but the answer in the long term is more supply. Besides accelerating the construction of new powerplants, there is the overhauling of the state-chartered independent electrical transmission grid operator (Cal-ISO) and the state-chartered wholesale power market, or power exchange (Cal-PX). Revising AB 1890 to allow a transition period for regulated utilities moving to a more deregulated world is also needed. The use of public dollars to finance development, private-public partnerships, or to capitalize publicly owned utilities are other issues.

New energy supplies are critical, with excess supplies, the market will self-correct. Without adequate supplies, the situation will worsen, the wholesale market will be stunted, and consumers will be in an artificial world of subsidized retail rates.

Two decades ago California promoted solar, wind, geothermal, biomass and small hydroelectric sources like no other state. An entire domestic renewable energy industry was initiated in California. Wind power was probably California's most famous renewable energy source and is now the fastest growing power source in the world.

The current regulatory system still encourages large, central station powerplants fueled by natural gas. We now have the diverse tools and technologies necessary to force a revolution in energy production that mimics, to a large extent, the evolution in scale and efficiency found in the telecommunications and computer industries. These newer technologies include solar photovoltaics, fuel cells and wind turbines which are the equivalent to wireless cell phones and portable laptops that replaced traditional grid-connected computers.

RENEWABLE ENERGY SOURCES

Fossil fuels such as petroleum, natural gas, coal, and nuclear fuels such as uranium must be discovered and extracted before they can be of benefit to mankind. This is a costly and time-consuming job. Also, the supply of these fuels is limited. There are other energy resources that are renewable. Their supply is virtually limitless or they can be replaced as needed.

Solar energy (energy produced from the sun) is available every day at sunrise. Rain replenishes the rivers and lakes that provide hydroelectric power (energy created by moving water). Winds that whip through mountain passes can be used to turn wind turbines for generating electricity on a regular basis.

Geothermal heat is heat created from molten earth deep below the surface. It can be captured and used to create steam to power turbines. Even wood products and garbage can be used as sources of fuel, as long as our forests and municipal wastes are managed properly.

Although renewable energy resources are available which can last indefinitely, several problems prevent their wider use as reliable energy sources. One is cost, collecting sunlight and converting it to electricity requires expensive solar collectors.

Additional technological developments are needed in several areas if renewable resources are to play a key role in energy. The reliability of wind turbines needs to be increased and their price lowered.

As the cost of fossil fuels continues to rise, the prospects of renewable energy resources will grow. New technology will reduce the price of energy from some renewable resources and make others easier to use.

HYDROELECTRIC POWER

Of all the renewable energy resources used in the United States since the 1930s, hydroelectric power (electricity generated by water-driven turbines) is the most important. During the 1930s, great dams were built across major waterways throughout the country, and nearly 40% of all U.S. electricity was generated by hydroelectric facilities. Today, water power accounts for only about 10% of America's total electric energy, although it is about 95% of all electricity generated from renewable resources.

The growth of hydroelectric power has slowed during the last few decades since many of the nation's most promising sites for dams are occupied. The Columbia River system has over 190 dams. The electricity generated by these and other dams costs approximately half as much as that generated from more traditional sources.

Environmental problems at each dam site include changes in reservoir water levels which affect plants and wildlife. Dams can lower downstream water temperatures when the cold water drawn from the bottom of the dams passes through the generators and is released downstream. The cold temperature can affect the life cycle of plants, insects, fish, birds and mammals.

If the licensing procedures for new dams could be eased, hydroelectric power could become an even larger source of renewable energy. Microhydro systems are also in use. Kitty Couch in the Black Mountains near Burnsville, North Carolina, has a microhydro system in a small, seasonal stream of 60-130 gallons per minute. The system generates about 5-kilowatt hours (kWh) per day in the dry season and about 10-kWh in the wet season. Figure 1-1 shows the basic elements for this type of system.

SOLAR POWER

After hydroelectric power, solar power is the most promising form of renewable energy in the United States today. Three areas of solar are prominent:

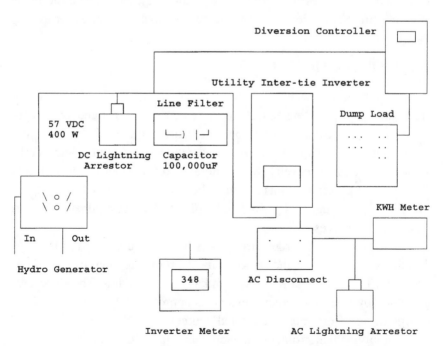

Figure 1-1. Grid-intertied Hydro Generator System

1. The heating and cooling of homes with solar energy.

2. The use of focused sunlight to create steam for generating electricity.

3. The direct conversion, using solar cells, of light into electric power.

Almost 25% of the U.S. energy is currently expended in the heating and cooling of buildings. One way using solar energy to heat a home is to use large windows facing south to capture the rays of the sun. Peter Berney in Prescott, Arizona, built his house in a U-shape surrounding a 400-square-foot greenhouse. The 16-inch poured adobe walls pass the heat at about an inch per hour into the house. Heat-absorbing walls can hold the heat for slow release into the home even after the sun goes down. Air ducts and fans circulate the heat to different rooms in the house.

Solar heating systems use black metal or plastic panels to absorb the sun's heat. Water circulates through the panels and transfers the heat into a building using pipes. Excess heat can be stored in the building as a tank of hot water. As heat is needed, a fan blows room air over the metal vanes attached to pipes through which solar-heated water is pumped from the storage tank.

Solar heating systems are effective but most homes using them require backup heaters that run on natural gas or electricity to provide heat on cloudy days or unusually cold nights. The backup heater adds to the initial cost of the home's overall heating system, although that cost is offset by reduced fuel bills.

One way to cool a house with solar energy is to use an electric heat pump. The pump works on a refrigeration cycle, cooling the air in a room by transferring heat from the room to the outside.

UNITS OF ENERGY AND POWER

There are many ways of measuring energy for the different forms of energy. One standard unit of energy is the same as the unit of work, the joule (abbreviated J), but a wide variety of other units are used too. Energy can also be measured in terms of quantities of fuel. Table 1-1 lists some common units and the conversion factors between them.

Table 1-1. Energy Units

---

1 kilowatt-hour (kWh) = 3.6 megajoules (MJ)
1 calorie = 4.18 J
1 Therm = 105.5 MJ
1 British Thermal Unit (Btu) = 1055 J
One horsepower (HP) = 746 W

Heat energy content of 1 cubic meter (m^3) of natural gas = 38 MJ
1 ton of oil = 12,000-kWh of electricity (assuming 100% conversion
   efficiency)
A large modern wind turbine produces about 500-kW
A large modern fossil fuel power station produces about 1-GW (1000-
   MW)

Power is usually measured in watts (abbreviated W), or in multiples of watts such as kilowatts or megawatts. One watt is equal to one joule per second.

Watts are familiar as ratings of electrical appliances. The electrical power is equal to the product of the voltage and the current at any instant:

$$Watts = Volts \times Amps$$

In an alternating current (AC) circuit, the voltage and current are not always in phase with each other and the power is given by:

$$Watts = Volts \times Amps \times Power\ Factor$$

Other, non-electrical energy conversion processes may also be described in terms of power.

Because of the close connection between energy and power, it is common to measure energy, and particularly electricity, in terms of power and time:

$$Energy = Power \times Time$$

Thus, a kilowatt-hour (kWh) is the energy of a 1-kW (1000W) device running for 1 hour (3600 seconds), and is equal to 3.6-MJ.

A battery uses stored chemical energy to provide its energy input to an electric circuit. A power station involves a longer series of conversions. If the input is a fuel, the first step is to burn it, using the heat to produce high pressure steam on hot gases. When the input is already in the form of a moving fluid such as wind or water, this stage is not needed.

The steam or other moving fluid is used to drive rotating turbines, which in turn drive the electrical generator. The generators operate on the principle that a voltage is induced in a wire that moves in a magnetic field. Connecting the ends of the wire to an electric circuit allows a current to flow.

The electrical energy can be transformed into heat, light or motion, depending upon what is connected to the circuit. The electricity becomes a convenient intermediary form of energy, used to allow energy released from one source to be converted to another different form some distance from the source.

Another form of electrical energy is carried by electromagnetic radiation. This electromagnetic energy is the form in which, solar energy reaches the earth. Electromagnetic energy is radiated in some amount by every object. It travels as a wave and can carry energy through empty space. The length of the wave is the wavelength and determines its form. This includes x-rays, ultraviolet and infrared radiation, microwaves, radio waves and the small band of wavelengths that are known as visible light.

Another basic form of energy is found in the central nuclei of atoms, and is called atomic or nuclear energy. The technology for releasing it was first developed for military purposes and then in a controlled version for the production of electricity. Nuclear power stations operate on the same principles as fossil fuel plants except that the heat from the fossil fuel is replaced by a heat from a controlled nuclear reactor.

CONVERSION AND EFFICIENCY

When energy is converted from one form to another, the useful output is never equal to the input. Some energy is always lost. The ratio of the useful output to the required input is called the efficiency of the process. Some of the inefficiencies can be reduced by good design, but others are inherent in the nature of the conversion.

Many systems involve the conversion of heat into mechanical or electrical energy. Heat is the kinetic energy of randomly-moving molecules, a chaotic form of energy. No machine can convert this chaos completely into the ordered state associated with mechanical or electrical energy.

This is the essence of the Second Law of Thermodynamics: that there is fundamentally a limit to the efficiency of any heat engine. Some energy must always be rejected as low-temperature heat.

Energy sources can be graded in quality. High-grade sources provide the most organized forms of energy and low-grade sources are the least organized. The higher grades include the kinetic energy of moving matter, gravitational potential energy and electrical energy. These can be converted with small losses. Larger losses occur when the lower forms are converted. Chemical energy occupies an intermediate position, then high-temperature heat and finally low-temperature heat.

About three times more heat is released at the power station than is delivered to the consumer. This wasted heat can be used for some other purpose. This is the principle of combined heat and power, (cogeneration) in which fuel is burned to generate electricity and the remaining output, which is in the form of heat, is used to provide low-grade heating. The overall efficiency of cogeneration plants can be 80% or more.

The waste heat from many high-temperature processes can be channeled to lower temperature use. This allows the energy of the original source to flow or cascade through two or more uses, thus providing more efficient use of fuel.

GENERATOR SYNCHRONIZING

The generators in modern power stations generate alternating current. The current increases and decreases sinusoidally a fixed number of times per second. This is 60 in the U.S. and 50 in Europe. The number of cycles per second, the frequency, is determined, in synchronous generators, by the speed of rotation of the generator, and is usually controlled within close limits. Many motors and appliances rely on a predictable frequency.

When many power stations are linked together by a high-voltage grid of power lines, it is necessary for all the generators to be synchronized, to operate at the same frequency, so that the peaks and troughs of current generated by each reinforce one another rather than canceling.

If generators driven by renewable sources such as wind or hydro power are connected to the same electricity grid, they too must be synchronized to all the others. The frequency of their output must be effectively constant, regardless of fluctuations in the renewable source.

This is a significant constraint on renewable sources, and can lead to lower efficiencies than could be obtained if the generators were allowed to vary in speed. In some renewable energy systems, such as wind turbines, asynchronous or induction generators are often used. The frequency of output of these generators is fixed by the frequency of the grid, but their rotation speed varies within a small percentage of the nominal speed, depending on the power output.

The steam turbine used in most fossil-fueled power stations is a heat engine and its efficiency is limited by the Second Law of Thermodynamics. This equation shows how to calculate the maximum theoretical efficiency. It is a function of the temperature of the steam reaching the turbine and the lower temperature at which the departing steam is condensed. If we call these T_U and T_L respectively, then:

$$\text{Maximum Theoretical Efficiency} = \frac{T_U - T_L}{T_U}$$

Kelvin temperature must be used in this formula. Suppose that steam enters the turbines at 550°C and is cooled to 27°C.

The Kelvin temperatures are $T_U = 823$ K and $T_L = 300$ K, so the maximum thermodynamic efficiency is 523/823 = 63.5%. In practice, older steam turbines achieve about 70% of this, or about 44.5% efficiency. Assuming that the efficiency of the boiler and generator is 90%, then the overall plant efficiency is 40%. Newer plants that use natural gas at very high temperatures in a gas turbine achieve a plant efficiency of close to 50%. The gas turbine is similar to an aircraft engine and uses some of its output gases to produce steam to run a conventional steam generator to produce electricity.

FOCUSED SUN POWER

Focused sunlight has been used to generate power in the deserts of southern California. One installation uses 130 acres with 1,818 mirrors. A motor controls each mirror, rotating it so that it reflects the sun's light to heat a tank of water. The reflected sunlight brings the water to a boil and the steam given off generates 10,000 kilowatts of electricity, enough to satisfy the needs of ten thousand people. The hot water can supply enough steam to operate the turbines without sunlight for up to four hours.

Electricity can also be generated from the sun's light using photo-voltaic, or light-sensitive, cells. Direct solar systems can produce several kilowatts of power.

In the Saudi Arabian desert, homes have been receiving power from a 350-kilowatt photovoltaic powerplant since 1982. In the United states, the first privately funded photovoltaic powerplant was built in 1983. It provides 1,000 kilowatts of power to the Southern California Edison Company.

Another private company built a 16,500-kilowatt photovoltaic powerplant in northern California. The Sacramento Municipal Utility District completed a 100,000-kilowatt plant.

WIND POWER

At the end of the 19th Century, almost 7 million windmills were supplying mechanical energy to pump water and grind grain in the United States. Wind power was vital to farmers on frontier lands.

Today, a new, more efficient type of wind machine is being used to help meet America's energy needs. These giant machines are much different from the small windmills that still pump water on some farms. Most modern wind turbines have only two or three blades, like those on an airplane propeller. Some are large enough to span a football field and can generate 5,000 kilowatts. There are home, marine and industrial wind turbine generating systems of several kilowatt capacity that can provide 40-kWh/month at 12 mph (5.4 m/s). In some parts of the country wind-powered generators can compete with oil-fired powerplants.

The National Aeronautics and Space Administration (NASA) has developed some of the world's most advanced wind turbines. The NASA MOD-2 design uses a 200-foot-high (60 m) tower with rotor blades some 150 feet (45-m) long. The height of the tower allows the unit to take advantage of strong winds, while the long blades are used to generate more power. A MOD-2 unit can produce 2,500 kilowatts of power in a 28-mile-an-hour (45-km-per-hour) wind. Winds of less than 14 miles (22.5-km) an hour are too light to generate power, and winds of more than 45 miles (72-km) an hour result in automatic shutdown of the unit to prevent damage.

Another design is the Darrieus windmill which uses thin blades connected at top and bottom to a vertical shaft. It resembles an egg beater and has several advantages over tower-supported wind generators. The Darrieus unit allows the heavy generator component to be placed on the ground rather than installed on a tower. The vertical blades of the Darrieus unit can accommodate winds from any direction, while propeller-blade turbines must be adjusted so that their blades face into the wind.

Ecological concerns may slow wind generator construction in some parts of the country. These giant moving blades can effect bird migratory patterns.

GEOTHERMAL POWER

Besides wind and water, there are other renewable energy sources. About 90 miles (145-km) north of San Francisco, CA, a series of hot springs (called The Geysers), has been used to produce steam to run turbines. This geothermal power station can generate almost 2 million kilowatts. The steam produced by The Geysers is known as dry steam because of its low moisture content. Dry steam is rare in the U.S., accounting for less than 1% of all geothermal resources. About half the states have hydrothermal, or underground hot water, resources. Some of these resources have been used to generate electricity, but many are still untapped.

Two methods of tapping geothermal power are used. When the geothermal hot-water temperatures are above 210°C (410°F), the direct flash method is used. When this superheated water is brought to the surface, its pressure drops and it starts to boil. The steam produced by the boiling water has been used to drive a turbine at a plant in Brawley, California. This plant generates more than 10,000 kilowatts of electricity. Geothermal sources could generate almost 25 million kilowatts in the U.S.

When the geothermal water temperature is below 210°C or the underground pressure is too low for the water to boil once it is brought to the surface, a more complex approach is used. The geothermal hot water is used to boil a second liquid with a much lower boiling temperature than water. The second liquid then vaporizes and drives a turbine to produce electrical power.

WOOD FUEL

Another form of renewable energy is wood and burnable refuse. The United States uses about 130 million tons of wood annually. These 130 million tons represent only a few percent of America's energy needs. Many homeowners burn wood in fireplaces and stoves to lower their heating bills, and some factories use wood scrap that was previously discarded to fire their furnaces.

New technology is helping to increase the efficiency of using wood as a fuel. Chipping machines allow the entire tree to be used. Pollution-control devices reduce the spread of toxic materials in wood smoke.

BIOMASS

One biomass conversion unit transforms wood chips into a methane rich gas that can be used in place of natural gas. Another biomass plant in Maine burns peat to produce power.

In addition to trees, some smaller plants, like the creosote bush, which grow in poor soil under dry conditions, can be used as sources of biomass, which are biological materials that can be used as fuel. These renewable sources of energy can be grown on otherwise unproductive land.

Another type of biomass used as fuel comes from distilleries using corn, sorghum, sugar beets and other organic products. It is ethyl alcohol, or ethanol which can be mixed in a ratio of 1 to 10 with gasoline to produce gasohol. The mash, or debris, left behind contains all the original protein and is used as a livestock feed supplement. A bushel of corn provides two and a half gallons of alcohol plus by-products that almost double the corn's value. Ethanol is a renewable source of energy, but critics question turning food-producing land into energy production.

WASTES

In addition to biomass, many communities around the country have used municipal wastes to produce energy. A plant in Connecticut burns garbage to produce electric power. Americans produce over 200 million tons of refuse each year. In the United States since 1960, the amount of solid wastes has grown by 80% and is expected to increase another 20% in the next decade. Some of this waste is industrial.

Every year, about 2 billion disposable pens are thrown away, along with 2 billion disposable razors and 16 billion disposable diapers. The major type of solid waste is paper and related materials, followed by glass, metals, food, plastics and yard refuse. All this waste is quickly filling the nation's landfills to capacity and disposal of all that refuse is an energy-consuming process.

Methods such as recycling, or turning used materials into new products, helps to conserve natural resources. Recycling paper eases the pressure on landfills and slows water pollution caused by paper mills while saving trees.

Recyling is costly but an alternative to recycling is to generate energy. When garbage is burned at 2,000°F (1,093°C) while air is forced into the mixture, most of the harmful by-products of incineration are eliminated.

The ashes make up only about 20% of the original volume of the garbage. The iron residues are removed with magnets and the remaining ash can be used on snow-and ice-covered roads instead of salt to limit environmental damage.

Another form of recycling is used by National Energy Associates in Imperial Valley, California. They burn dried cow manure, converting 900 tons a day into 15,000 kilowatts of power.

Another powerplant burns wheat straw and other crop wastes. This plant, was constructed by John Hancock Mutual Life and Niagara Mohawk Power, who were partners in the venture.

Another source of renewable energy is currently being used in Europe. It is called tidal energy, and it involves the natural rhythm of oceanic tides. A tidal powerplant built in 1966 on the Rance River in Brittany, France, uses the movement of ocean tides to run 24 turbines at the river's mouth. The plant generates 240,000 kilowatts of power at peak water flow, but it can operate only about 1/4th of the time because of the changing currents.

The percentage of energy generated from wood, biomass, manure, refuse, and the oceans' tides remains small, but as technology improves to squeeze more energy from these forms of fuel, their use is expected to grow.

DEREGULATION

So far, 24 states and the District of Columbia have instituted electricity deregulation. In almost every case the process has had some bumps along the way, largely because deregulation is being attempted during a period of short power supplies and escalating demand. Some states are hurting less than others because utilities, regulators, and lawmakers have done a better job adjusting to problems.

A critical issue is ensuring adequate supply, when prices began to rise after Illinois deregulated in 1997, the state began building new powerplants. To speed up the process, its largest utility, Commonwealth Edison Company, promoted a list of sites in rural or industrial areas that wanted the jobs or property taxes that the new plants would bring. The Illinois Corporate Commission, which regulates utilities, estimates that new plants are now producing enough new electricity to supply almost 4 million homes.

In California, utilities and power producers had to buy and sell electricity on a last-minute daily basis through a pooled entity. Pennsylvania allowed its utilities to buy power from whomever they wanted. In Pittsburgh the Duquesne Light Company signed long-term power contracts that cut consumers' power bills by 20%.

New York, like California, had utilities sell many of their powerplants as part of deregulation. But, unlike California, utilities were allowed to pass rising fuel costs to consumers.

California has used the British deregulation model which also ran into problems early, but it is now working much better. Prices jumped in the early 1990s and wholesale price caps were implemented. Later, windfall-profit taxes were placed on power suppliers. Utilities were forced to cut their employees by half. The country power pool which is similar to California's Power Exchange was revamped to encourage longer-term contrasts. Retail power prices have declined 30% in England over the past decade.

In the U.S., over the last decade, power surpluses resulted from the overbuilding of large centralized nuclear and coal powerplants in the 1970s. This, in turn, depressed electricity prices. Prices are now escalating due to a supply shortage. For the first time in several decades, electricity consumption is going up, not down. In Silicon Valley, electricity consumption has been growing at 5% annually over the past few years. While electricity represented only 25% of our total energy needs 25 years ago, it will represent half of our energy consumption in a few more decades.

One study by the Lawrence Berkeley National Laboratory in California focused on electronic products that continue to use power even when turned off. The study determined that the nation wastes over 70-terawatt hours (tWh) per year on such equipment. This is about 2% of the total power production.

About 75% of this waste occurs in commercial offices, 12% in home offices and the rest in industrial sites. Desktop computers and monitors are the biggest offenders and waste over 14-tWh each. Copiers waste another 7-tWh and laser printers are close behind with 6-tWh. Fax machines and inkjet printers waste another 3-tWh each.

NATURAL GAS

Almost all new powerplants burn natural gas. Natural gas is formed in the same deposits as oil and by the same natural forces. Natural gas did not become popular until after World War II when economical techniques for transporting it were discovered. From the mid-1940's to the mid-1960's, U.S. consumption of natural gas nearly quadrupled as it proved to be a cheap, clean-burning fuel. During this 20 year period, the price of natural gas was regulated by the Federal Power Commission (FPC), which made it a good buy relative to other fuels.

Since it is the cleanest of fossil fuels, natural gas has become the fuel of choice for generating electricity. Since 95% of all newly proposed powerplants, some 250,000 megawatts run on natural gas, supplies have dwindled. High natural gas prices may last for at least several more years because of supply shortages.

Millions of consumers have gotten reminders of how free market forces work, in the form of unusually large natural gas bills in their mailboxes. U.S. Department of Energy estimates showed that heating an average home in the Midwest would cost 54% than the year before. In the Washington, DC area, prices have more than doubled. In the past few years natural gas prices have been rising. A few years ago natural gas prices hit bottom and these low prices led to a cutback in the exploration and drilling of new gas supplies.

Since it takes time for new supplies to emerge, only now are gas supplies beginning to increase. In 1999, about 400 natural gas rigs were in operation. That number has doubled.

Homes with large gas bills have cut back on consumption. High prices encourage supply while limiting demand, and eventually bring prices down.

There was not a natural-gas supply shortage. When prices spiked, gas companies did not have to cut back on service to consumers. Contrast this to the electricity situation in California, with price controls and blackouts. Natural gas consumers face the prospect of high prices in the future, due to government policies that push up demand while keeping off limits vast supplies of natural gas that could meet this demand.

NATURAL GAS DEMAND

Government has encouraged the use of natural gas for two reasons. Almost all of the natural gas the United States consumes is produced domestically. Only about 15% is imported, and most of that comes from

Canada. This situation is unlikely to change. Unlike oil, natural gas cannot be transported easily over the ocean since it must be liquefied first. Domestic supply and supplies from nearby countries is tapped from underground reservoirs piped to users along a vast array network of underground pipelines. Encouraging the use of natural gas helps cut the dependence on oil, almost two-thirds of which is imported.

Another reason natural gas is popular is that, compared with other major sources of energy, natural gas is far less harmful to the environment. It produces about half the nitrogen oxides of oil, per unit of electricity generated, for example. It emits almost no sulfur dioxide or soot.

Advanced natural-gas cars cut carbon monoxide emissions by 80% compared with conventional, gasoline-powered vehicles. Natural gas produces about half as much carbon dioxide for each unit of energy as coal, and almost a third less than oil.

Most projections envision a dramatic rise in natural gas consumption which will require substantial increases in natural gas development and production to meet the growing market demand.

Since natural gas is cheaper and more efficient in many cases than alternative fuels, the country's use of natural gas has been climbing sharply. Over the past decade, gas consumption has increased by 12%. The country uses over 20 trillion cubic feet (tcf) of natural gas per year. This is expected to climb to at least 30 tcf over the next decade. Over 95% of all new electricity plants are powered by natural gas, and about 70% of new homes are heated by it. Cities are using natural-gas-powered bus and government-owned car fleets.

LIMITING THE SUPPLY

The problem is that while the government has been pushing natural gas demand, it has also been keeping vast stocks of it locked up underground because of restrictions on what can be done on the land and sea above.

In a 1999 report, the National Petroleum Council, which advises the Energy Department, stated that over 200 trillion cubic feet of natural gas are off limits to drillers due to regulations in drilling. In the Rocky Mountain region, over 130 tcf are off limits in an area which includes most of Colorado, Wyoming and Utah, and parts of Arizona, New Mexico, Kansas, Nebraska, South Dakota and Montana.

In addition, over 50 tcf of natural gas are located off shore on the East and West Coasts, none of which can be drilled because of bans in those areas. Over 20 tcf are blocked because of bans on drilling in the Gulf of Mexico.

In the Rocky Mountain region, there are a wide range of wildlife restrictions during various times of the year to protect such animals as prairie dogs, raptors, burrowing owls, sage grouse, and big game. These actions taken together effectively prevent access to natural resources such as natural gas. More supplies were taken away through national monument declarations and U.S. Forest Service plans to bar road building in millions of acres of forest. That act alone put 10 trillion cubic feet of natural gas out of reach.

Drilling can be disruptive, but technological advances over the past decades have sharply reduced the harm to the surrounding environment. According to the Department of Energy these advances include horizontal and directional drilling that let drillers mount a rig in one area and retrieve supplies miles away. There are also better detection methods using satellites, remote sensing, and super computers that reduce disruption of the land and produce fewer dry holes resulting in less waste.

The Energy Department concluded that resources under arctic regions, coastal and deep offshore waters, wetlands and wildlife habitats, public lands, and even cities and airports can be extracted without disrupting surface features above them.

Canada has made use of these technological advances. An offshore gas rig off the coast of Nova Scotia provides gas to New England. It is just north of the American coast where drilling is forbidden. Canada's natural resources department decides new parkland boundaries only after an assessment of minerals underneath is conducted.

A few years ago Canada passed the Oceans Act, which provided integrated coastal zone management. Canadian officials assess the area and come up with a consensus on how industries and other groups with interests in the region can coexist.

In the U.S. a similar approach is needed, one that tries to balance the task of supplying the nation's growing energy needs with preserving and protecting the environment.

As older coal, oil and natural gas plants run more often in supply shortage emergencies, pollution goes up. But Internet era companies need the super-reliability that the older power grid-connected powerplants cannot always supply. Many of these companies are installing diesel generators on site, a trend that threatens to increase air pollution, testing the belief that e-commerce is an environmentally friendly development.

The primary emphasis of power growth has been in natural gas plants, by 2001 the California Energy Commission approved 4,000 megawatts with another 24,000 megawatts waiting in pending applications.

POWERPLANT PROBLEMS

Most of today's electricity is generated in steam-turbine powerplants that use coal, oil, natural gas or nuclear reactors. The steam is heated at high pressures to more than 1,000°F. It passes through a turbine where it is cooled to less than 100°F. One turbine-generator can supply the needs of about 500,000 homes.

There are severe problems to the successful siting of gas-fired powerplants. Poor air quality often exists in regions where new supplies are most needed. New natural gas powerplants are relatively clean, but they still emit pollutants.

In order to site these plants, the developers need to reduce pollution elsewhere. This is often done typically in excess of what is generated at the new powerplant by buying so-called offsets. In these transactions, the powerplant developer pays other air pollution sources to reduce their emissions. These offsets are becoming quite scarce and their costs have jumped due to the short supply.

Powerplants use a lot of water for cooling, and there is opposition from other water users to diverting this resource to powerplants. Some gas-fired plants can use dry cooling, but that raises the price. The cost of electricity from gas plants has already more than doubled recently because of fuel price increases and a dry cooling system add another layer of costs.

The biggest barrier to adding natural gas powerplants may be the availability of the fuel itself. The natural gas may be in the ground, but it takes time and money to get it out and to the plants. Since natural gas is also a primary heating fuel, supplies can be expected to be short.

The constraints on shipping additional natural gas supplies into California are matched by a lack of transmission access to import additional electricity from powerplants burning coal or splitting atoms located in other states. The uncertainties associated with an industry that is partly regulated and partly competitive makes it difficult to invest in needed infrastructure, particularly transmission facilities.

California's deregulation plan did not create a mechanism for private, non-utility investment in power transmission. Utility owners of the transmission system have no incentive to invest in transmission.

The transmission system erected in an era of vertical integration and was never meant to handle the loads of an unbundled electricity sector, where power producers supply bulk electricity to the highest bidders. Transmission will require immense investment in the coming years in order to maintain system-wide reliability. But, investments in transmission facilities have gradually declined over the last 20 years.

Innovation is needed at the distribution level of utility service. Cutting edge distributed generation sources could be integrated into grids to address the challenges posed by the current electricity infrastructure.

Reliability is a major concern for Internet-related businesses. A power outage on one day is estimated to cost $100 million in lost e-business. Oracle puts the price of a power disruption at millions of dollars per hour. Hewlett Packard estimates that a 20-minute outage at a circuit fabrication plant can result in a $30 million debt due to the loss of a day's production. As the power grid becomes less reliable, the case for innovative supply solutions becomes stronger. These high-availability power supplies include non-polluting fuel cells. More kilowatts may not be enough, there is also a need to have reliable kilowatts, to meet high tech needs.

FUEL CELLS

Several companies, including Pennsylvania Power & Light and Toshiba, announced major investments in fuel cell technology after the price spikes in electricity. Almost $17 billion of the $20 billion invested in alternative energy technologies has gone to fuel cell companies. Some fuel cell stocks have jumped as much as 1,000 percent.

Fuel cells rely upon an electrochemical process to convert chemical energy into electricity and hot water. They can be used to generate power during blackouts. Fuel cells can run on a variety of fuels that include natural gas and hydrogen. A 1-megawatt system is used to provide power for a post office in Anchorage, Alaska. The Los Angles Department of Water and Power has installed a fuel cell system. It believes these clean distributed generation sources are key solutions to California's power supply problem.

Diesel generators are the most common form of distributed generation in operation. These can help fill in supply gaps but air quality is impacted since diesel is viewed as a dirty technology, particularly when there are other alternatives that can deliver power reliability with less impact to the environment.

Leading companies such as Cisco Systems provide computer networks that are dependent upon ultra-reliable electricity. Cisco has installed 12-megawatts of back-up diesel generators. Diesel generators are 10 times more polluting than the vintage natural gas plants built during the Korean War that, in turn, are 10 times more polluting than the large gas plants currently being built.

Fuel cells can generate clean electricity from a variety of fuels including hydrogen and could supply as much as 20% of the nation's commercial and industrial power needs within a decade. Solar photovoltaic technologies currently dominate clean energy distributed generation purchases by businesses and other consumers.

SOLAR PHOTOVOLTAICS POWER

PowerLight Corporation is a solar photovoltaics manufacturer that views power price spikes in California as a business opportunity. Installation of 1-megawatt of solar power on large commercial buildings is possible. The price of photovoltaic power is high but within in the broad range of 10 to 35 cents per kilowatt which is about the same or lower than being paid for dirty coal power during supply emergencies.

Solar photovoltaic panels can be installed in a matter of days which is far quicker than the years it takes to site and build most large fossil fuel facilities. Germany and Japan install roughly 70 megawatts of photovoltaic per year.

PowerLight installed a 100-kilowatt system on the top of the Anaheim Convention Center in 2000. PowerLight estimates that California could generate 16,000 megawatts of electricity from the sun if every commercial and industrial roof were covered with photovoltaic tiles. This amount would represent about a third of the state's entire demand for electricity. In California solar photovoltaics are being spearheaded by local governments.

The National Renewable Energy Laboratory has studied the potential benefits of installing solar photovoltaic in our nation's urban centers. Seven major outages, including one impacting San Francisco, were analyzed from the aspect of the quality of solar energy during the exact times of the power loss. In all but one of the outages, the conditions for optimal solar electricity generation were about 90%. This is because the sun is strongest on days that lead to the heat waves that stress our electricity delivery system. We can use the same sun that helped create the crisis to solve the power supply problem.

A study by the U.S. Department of Energy showed that solar photovoltaic panels on the roofs of California's city and county buildings could generate 200 megawatts of clean electricity. School roofs covered with solar photovoltaics could add another 1,500 megawatts to the state's peak power supply.

Proposals for distributed power sources have encountered extended delays, often due to local governments. Many local government officials are not up to speed when it comes to these micro-powerplants.

In the early 1980s, when fuel cells first arrived, utilities were installing them and there were few problems regarding the codes and regulations that applied at the customer site. Today, typically a third party installs fuel cells at customer sites and local government approvals are rocky. Local officials need to understand the technology better.

To most local building departments, a fuel cell is just a big box, and they are not sure what to do when it comes to issuing a permit for construction or operation. Like solar photovoltaic panels, fuel cells do not release any harmful air pollutants because nothing is burned. But, they do produce high-temperature steam, which can reach 1,000°C although the potential for problems is considered minimal. Many of the local codes do not address these technologies at all. Some argue that they are an appliance but others argue that there are more serious impacts on health and safety.

RealEnergy supplies solar photovoltaics and is mounting an acre of solar panels on the roofs of the City Center in Los Angeles, which is owned by Arden Realty. RealEnergy is working with major owners of commercial property taking a long-term view on electricity reliability and environmental impact. Arden Realty's goals are to install equipment that increases the reliability of power systems for tenants and reduces the environmental impacts of dirty electrical generation. There will be more than 300 kilowatts of solar photovoltaics on the City Center which makes it the largest private installation of this technology in the Western Hemisphere.

Solar photovoltaic installation involves high up-front expenditures, but RealEnergy believes that the costs are reasonable when spread out over the 20 year life of the equipment. After the installation, there is no fuel and little maintenance.

The California Energy Commission included buy-down funds in the state's deregulation law which can cover almost half of a new solar photovoltaic installation. This has helped to boost demand for grid-connected solar systems by a factor of ten.

In California, electricity marketers and generators were accused of profiteering and market manipulation. A trading company produces little or no power itself and owns relatively little in the way of hard assets. Instead it has pioneered the financialization of energy. Its profit stream comes from a flow of often low-margin trades, in which it buys and sells a variety of contracts. Beyond energy, these new market models include data storage, steel and even advertising space.

Power deregulation has lost momentum since it is hard to argue that it has resulted in lower bills. In overseas markets, big utilities in the south of Europe, want to use the California experience to slow down power liberalization according to the European Federation of Energy Traders.

The power brokers believe that the contention in California has buried the reasons for deregulation. The older utilities were expensive to operate and provided often appalling service to their customers. In the regulated model, utilities could pass on costs to customers and they lacked incentives to utilize capacity more efficiently or to offer innovative service.

Oregon, Arkansas, and Nevada are considering a slowdown in their deregulation plans, while northeastern states have put in place wholesale price caps.

A large trading company may have over 1,000 traders making markets in gas, electricity, metals, bandwidth, and other products. They will post prices for an array of energy contracts. Utilities caught short on supply can buy more and generators with excess capacity can find markets willing to pay.

In California, business leaders pushed deregulation because they were paying 50% more for power than their competitors in other parts of the country. The utilities were operating high-cost plants with little access to capital to upgrade. The upside of deregulation has been the huge amount of new construction. It is the direct result of the industry having new access to capital markets.

One problem now is that the nation is in a confused state between regulation and deregulation. The freezing of retail prices for consumers and businesses in California while deregulating the wholesale market where utilities shop left the power companies unable to cover their rising energy costs and pushed them to financial collapse. A professor at the Yale School of Management, compares this to the blunder that regulators made in the 1970s with the savings and loan industry. The S&Ls were restricted to investing in long-term assets, like mortgages, while paying volatile market rates on shorter-term deposits.

ENERGY PRICING

For a long time, energy production costs were low and these production costs were reflected in the low energy prices charged to consumers, who had little or no incentive to limit their energy consumption. Today, increased prices to the consumer are an important incentive to encouraging conservation.

Energy has traditionally been priced to reflect the average cost of old, cheaper energy supplies or plants and new, expensive supplies or plants. While renewable energies are free, the equipment needed to capture conventional energy resources. But, renewable energies are often competitive when the life-cycle costs are computed and compared. This is the total cost of equipment plus fuel over a system's lifetime.

RATE STRUCTURES

The prices charged to consumers for electricity and natural gas reflect a good deal more than the cost of energy. These rates, which are generally established by the public utilities commission in each state, traditionally have been based on various factors in addition to fuel costs. These factors include the following:

1. Costs of building energy production and distribution facilities.

2. The allocation of costs among different types of consumers based upon consumption amounts and patterns.

3. An allowance for a reasonable rate of return to the utility.

4. The recognition of rapid fuel price increases, primarily through fuel adjustment charges.

In the past, rate structures have generally been designed to encourage consumption rather than conservation. In the future, energy consumption up to a certain average level could be charged at current or slightly higher rates while any energy use in excess of this level would be charged at a higher rate. Under this plan consumers would not be penalized for moderate energy use but would pay more for increased or excessive use.

Flat Rates

When you buy oil or gasoline, you generally pay a fixed price per gallon. This neither encourages nor discourages conservation as long the price remains constant. Residential consumers of less than 500- to 600-kWh of electricity per month often pay similar flat rates. These rates usually reflect the average cost of electricity. A flat rate also remains constant throughout the day.

Declining Rates

Gas and electricity have traditionally been sold at declining rates, under which each therm or kWh is cheaper than the last, in order to encourage consumption. This is still the practice in many areas. Declining rates are constant throughout the day. Since it is difficult in practice to charge a little less for each successive kWh consumed, declining block rates are normally used. For very large consumers, electricity can be extremely cheap. Typically, for the first 800-kWh consumed, the rate would be the highest per kWh, for the next 4,200 kWh the rate might be 30% less and for anything above 5,000 kWh another 30-40% less. The average cost of electricity for 10,000-kWh of consumption would be less than half per kWh compared to the initial block.

Inverted Rates

In inverted rates, each kWh consumed is more expensive than the previous one. This type of rate structure encourages efforts to limit consumption. An inverted rate remains constant throughout the day.

In practice, increasing block rates are more likely to be used. The first block of energy is relatively cheap, while each increasing block gets more expensive. The first 800-kWh would be at the lowest rate, the next 400-kWh would cost 30% more per kWh and anything above 1,200-kWh could cost twice the initial rate. This is known as an inverted block rate structure. Life line rates could be used in which the cost of an initial block of energy is low, with the cost increasing for increased consumption above the initial block. The initial block might be 400-kWh. Above that, electricity costs would double or triple per kWh.

Demand Charges

Each consumer requires a certain transmission and powerplant capacity to supply energy at their rate of consumption. The gas or electric utility must invest money to build this capacity as well as to purchase the fuel to meet that need. Some consumers have an almost constant need for energy while others have a high periodic or occasional consumption. In order to meet the customer's peak demand, especially if it is coincident in time with the utility's daily peak, the utility may impose an added charge on consumers to pay for the extra capacity. This also acts as an incentive to limit demand peaks. In a large building with a single meter, the demand charge is typically 1/3 of the total bill.

This provides an incentive to reduce peak demand through load management. The demand charge is usually a constant dollar figure charged for each kilowatt of a peak demand and could be a few dollars per kilowatt. This charge may increase as peak demand increases.

Time-of-use Rates (TOURS)

This type of rate includes a variable demand charge or energy price that changes with the time of day. Just as long-distance telephone charges vary during the day, time-of-use-rates reflect the changes in hourly energy demand. Such a rate provides an incentive to shift energy consumption to those times of the day when demand is lower and energy is less expensive.

Interruptible Rates

An extreme application of TOURS is a rate that allows the utility to cut off part or all of the customer's demand when the utility chooses. The utility is allowed to do this for limited number of times per month and for a limited duration. In return, the consumer receives a major discount for what is a lower-than-normal standard of service. This allows the utility to avoid the need to construct additional capacity to meet peak demand.

Industry's response to the need to reduce energy use falls into three stages. During the 1970s, a crisis atmosphere was apparent. Plants emphasized turning down and turning off. In the 1980s, technological advancements led to the replacement of obsolete equipment and the installation of microprocessor-based devices. In the 1990s, energy and environmental issues merged together, and energy conservation efforts are being refined into management processes.

A common misconception is to consider energy reduction simply as a energy conservation program. Energy is conserved by turning off equipment and shutting down processes. But, these measures have a negative impact on plant production. Energy needs must be managed to produce the final product or service at the minimum consumption.

The goals of a successful energy management process include maximizing producer efficiency while minimizing consumer energy use. You also want to maintain a high energy load factor and use energy in the most economical way.

The plant utilities must always be adequate to meet production/process demands. If the output of the plant is threatened because energy supplies are inadequate, accommodations must be made so that production/process targets are met until a new energy supply or source is developed.

ENERGY MANAGEMENT PROGRAMS

Energy management programs provide a challenge to create an energy awareness and provide tools for reducing energy consumption to the low levels achieved during the emergency situation. A structured approach to energy management focuses on the following guidelines.

1.   Total management commitment including the support of the company president as the first step.

2.   Employee cooperation and input is vital.

3.   Energy surveys are critical but are often given too little time and effort.

4.   Analysis of survey results, often too little time is devoted to analyzing where and why energy is used.

5.   Conservation goals are absolutely necessary. Without goals, there is nothing to strive for and no method for measuring performance.

6.   Reporting and communication is vital to energy management.

7.   Implement changes. Activities should be included whether it be disconnecting excess light fixtures or adding computer-based enthalpy controls on air washers to make use of outside air.

8. Provide the necessary equipment. This could be sensors and computers to log temperature and kilowatt demand with simple data logging programs.

9. Monitor results. Problem situations may return to their original state unless they are monitored continuously. Continuous monitoring should be maintained.

ENERGY COORDINATORS

The success of the energy management process depends on the response of plant personnel. One way to guide the effort is to appoint plant energy coordinators. Coordinators should have the following responsibilities:

*   Stimulating interest in energy projects,

*   Providing operators, foremen, and resource people with the tools they need for energy management, and

*   Making energy management an integral part of every department.

A coordinator should be a leader and understand general engineering and plant operations. They should be trained in energy management. Seminars or other educational classes can be used. The functions of a coordinator include:

*   Inform plant management of any problems in energy use reduction and suggest ways to solve them,

*   Ensure that cost reduction measures are on schedule,

*   Establish priorities, set completion dates, and obtaining adequate staffing,

*   Develop new energy reduction projects,

*   Determine trends in energy usage and efficiency,

*   Develop checklists to improve efficiency and operations techniques, and

*   Develop energy standards.

The energy management program should be outlined so all personnel understand the objectives and their role in them. The objective might be to ensure continued energy availability at the lowest possible unit cost.

Energy should be available in sufficient quantity to support production/processing. If energy sources have historically been inadequate, alternatives should be investigated.

Efforts to reduce Btus used per unit made should be directed at controlling cost per Btu by selecting from among available energy sources. Capital and operating expenditures and human resource requirements should be managed to yield the best return on assets.

Disseminating information about the energy management program should include circulating energy use and cost information, displaying posters on bulletin boards and conducting seminars.

A critical part of controlling energy use is to develop and apply accurate energy standards. Standards may be variable, those associated with production, fixed, that referred to as services energy. Energy use is dynamic and efforts to improve standards should be ongoing so that changes in product or service volume and mix, changes in the efficiency of utility equipment and seasonal fluctuations can be accurately modeled.

ENERGY AUDITS

Energy auditing is another part of controlling energy use. An audit can point out when equipment is operating unnecessarily or wastefully. It helps to develop operating procedures for reducing energy waste and can quantify the cost of that waste. Auditing also allows priorities to be set and helps to identify problems that require new procedures to reduce costs. Auditing can determine energy consumption rates and costs for the equipment in each department. The energy coordinator should be able to provide specific cost data.

Department level energy coordinators should be able to identify major energy users and consumption rates. They should also be able to develop procedures that reduce waste and identify problems. Waste reduction may require process or equipment modifications.

Department energy coordinators can determine how long equipment is in service without performing a useful task. This should be done frequently enough to obtain statistically valid data. Checklists and sampling procedures can assist in this procedure.

The cost of energy wasted is the product of the length of time energy is wasted and the hourly or daily cost to operate the equipment. This is illustrated in Table 1-2 for some common system loads.

Table 1-2. Typical System Loads

Load	Watts	Average Hours/Day	Average WH/Day
Induction cooktop	1,600	1	1,600
Well pump	250	6	1,500
Washing machine	800	1	800
Lights	200	4	800
Convection oven	1,500	0.5	750
Band saw	1,450	0.29	414.3
Vacuum cleaner	1,000	0.5	500
H_2O conditioner	10	24.00	240
Toaster, four-slice	1,800	.1	180
Clock	4	24	96
TV	80	1	80
Computer monitor	100	.5	50
Dishwasher	1,400	.1	140
Fan	600	1	600
Iron	1,200	.1	120
VCR	30	1.0	30
Lathe	900	0.03	30.0
Computer	55	0.50	27.5
Battery chargers	160	0.14	22.9
Stereo	30	0.33	10
Ceiling fans	30	1	30
Drill press	800	0.01	5.7
Air compressor	1,000	0.01	5.6
Printer	35	0.07	2.5
Coffee grinder	100	0.02	2.4

Approximately two to eight man-weeks may be needed to complete an initial audit in a typical plant. The actual time is influenced by equipment complexity, forms of energy used, and quality of available drawings.

Most industries can justify one man-year for each $1 million spent annually on energy. As the program progresses and savings are achieved, the effort can be down-sized to one man-year for each $2 to $5 million spent each year.

Some plants start with a small effort, working on simple projects with fast paybacks. As the projects become more complex and investments increase, more resources are allocated.

References

Asmus, Peter, "California's New Energy Legacy: A Desperate Innovator?" *California Journal*, Vol. XXXII No. 1, January 2001, pp. 17-21.

Boyle, Godfrey, Editor, *Renewable Energy*, Oxford, England: Oxford University Press, 1996.

Eisenberg, Daniel, "Which State is Next?," *Time*, Vol. 157 No. 4, January 29, 2001, pp. 45-48.

Greenwald, John, "The New Energy Crunch," *Time*, Vol. 157 No. 4, January 29, 2001, pp. 37-44.

Herda, D.J. and Margaret L. Madden, *Energy Resources*, New York: Science/Technology Society, 1991.

Meline, John W., "Why Natural Gas Problems Loom," *Consumer Research*, Vol. 84 No. 3, March 2001, pp. 10-14.

Palmeri, Christopher, "California: It Didn't Have to be This Way," *Business Week*, Vol. 1, January 22, 2001, p. 40.

"Power Play," *Business Week*, Vol. 2 No 2, February 12, 2001, pp. 72-80.

Scientific Staff of the Massachusetts Audubon Society, The Energy Saver's Handbook, Emmaus, PA: Rodale Press, 1982.

Internet: www.web5.infotrac.galgroup.com

Stibbens, Wayne L., "Keeping the Energy Management Process Alive," *Plant Engineering*, Vol. 47 #10, June 3, 1993, pp. 83-84.

CHAPTER 2

ENERGY SOURCES— HEAT PUMPS, SOLAR ENERGY AND WIND POWER

HEAT PUMPS

A heat pump operates like a refrigerator or an air conditioner and removes heat from one area called the source and transfers it to another called the sink. Heat pumps can be used for both heating and cooling. Electric heat pumps can deliver up to three times more heat energy than is contained in the electricity used to power the device. Including efficiency losses at the electric powerplant, from 75 to 100% of the heat energy in the original fuel can be delivered by an electric heat pump.

When a substance changes from a liquid to a gas by boiling or evaporation, it absorbs heat. When it changes back from a gas to a liquid by condensation or liquefaction, it releases heat. A heat pump uses these properties to transfer heat. The refrigerant has a boiling temperature far below room or outside temperatures. It is contained in a liquid state in a sealed coil, the heat exchanger, and exposed to a heat source. This may be outside air, solar-heated water or groundwater.

The refrigerant absorbs heat from the source and, as it warms up, it boils changing to a gas. The refrigerant is then pumped through a compress or into another sealed coil which is exposed to the sink. The compressor increases the pressure on the vapor, causing it to condense. As the refrigerant condenses, it releases heat which the sink absorbs. The sink is usually indoor air or a hot water heating system. Thus, heat is removed from the source and transferred to the sink and the cycle can be repeated indefinitely.

A heat pump can be operated in reverse where the source is indoor air and the sink, outdoor air. This has the effect of cooling, rather than heating. Cooling heat pumps are known as air conditioners. Central heat pump systems are ordinarily used for both heating and cooling. Along with the heat extracted directly from the source, heat pumps are able to use the waste heat produced by the compressor motor. The total amount of heat delivered by the heat pump to the sink equals the amount of heat extracted from the source plus the waste heat from the compressor motor.

The efficiency of a heat pump is given by the ratio of the heat delivered to the sink to the heat content of the electricity or other power used by the compressor motor. This ratio is called the Coefficient of Performance (COP). Electric heat pumps normally have a COP of 2 to 3 compared to a COP of 1 for electric resistance heating.

Counting the efficiency losses at the electric powerplant, a heat pump using electricity produced by the burning of oil or gas with a COP of 2.5 provides about the same amount of heat as the direct burning of fossil fuel with an efficiency of 75%. The COP of a heat pump decreases as the temperature of the source decreases since more work is required to supply the same amount of heat and more fuel is consumed. An electric heat pump with a COP of 2 at a source temperature of 45°F may only have a COP of 1.3 at 25°F. This makes electric heat pumps more expensive than electric resistance heating at source temperatures below about 20°F. This occurs in northern climates for heat pumps that use outside air as a source. Most electric heat pumps use electric resistance auxiliary heaters that switch on when the outdoor temperature drops below 20°F. This problem does not exist if the source is groundwater since, even under most conditions, groundwater temperatures rarely drop below 40°F.

HEAT SOURCES FOR HEAT PUMPS

Heat pumps can extract heat from a variety of sources. When the cost of electricity is low and winter temperatures are in the 35 to 55°F range, electric heat pumps can be competitive with fossil fuels. When temperatures average 30°F or less and electricity becomes more expensive, air-source heat pumps become more expensive to operate than oil or gas systems. Groundwater is one way to increase the efficiency of a heat pump system in a colder climate. This provides a moderately high and constant temperature heat source.

Wells can supply heat source water to several buildings. Heat pumps can also be operated in tandem with an active solar heating system. The solar system provides warm source water for the heat pump. On clear days, the solar system can displace the heat pump or boost the temperature of the storage tank. On overcast days, the heat pump can compensate for the low solar system efficiency. In the summer, the solar system can supply domestic hot water. The ground is a potential heat source. Ground temperatures below the frost line are generally the same as groundwater temperatures. Coils of pipe buried below this depth can extract heat directly from the ground. About 1,000 square feet of ground space are required as a heat source for a single-family dwelling.

Heat pumps can also use waste heat from other energy-consuming systems as a source. Large buildings or complexes, such as hospitals often dump their waste heat into the air. Some electric utilities offer rates that vary during the day. These rates provide low nighttime electricity costs so it is possible to use the heat pump to heat water at night, circulate the hot water through the distribution system during the day and save on energy costs over fossil fuel systems. Heat pump capacity is the rate of heating or cooling supplied at a specific temperature, usually around 45 or 50°F. It is expressed in thousands of Btu's per hour (KBh) or tons where 1 ton = 12,000 Btu per hour. On a winter day of 30°F, the actual heating capacity may be much lower than the advertised capacity. Heat pumps are often used primarily for air conditioning in warmer climates and are often sized to meet the cooling load. Four to six tons is a common size for an uninsulated single family home. If solar heating is impractical and you have electric resistance heating, a heat pump makes sense. If time-of-use electric rates are in effect, heat pumps with short-term heat storage capacity may provide heat more cheaply than fossil fuel systems. Icemaking heat pumps can make ice for air conditioning when electricity costs are low. The system can also heat domestic hot water with reject heat.

PASSIVE SOLAR HEATING

Passive solar heating involves the use of the building construction in order to capture and store solar energy without requiring electrical or mechanical devices such as pumps or fans. Since there are few moving parts in passive systems, system maintenance is minimal. The operation usually requires some degree of control to effectively manage the daily heat gain and minimize the nighttime heat loss. There may also be backup heating. Even on the coldest sunny days, a passive heating system can supply a significant portion of a building's heating load. Passive solar systems can also collect heat on overcast days.

A source temperature of 45°F may only have a COP of 1.3 at 25°F. This makes electric heat pumps more expensive than electric resistance heating at source temperatures below about 20°F. This occurs in northern climates for heat pumps that use outside air as a source. Most electric heat pumps use electric resistance auxiliary heaters that switch on when the outdoor temperature drops below 20°F. This problem does not exist if the source is groundwater since, even under most conditions, groundwater temperatures rarely drop below 40°F.

SOLAR COLLECTION

Solar collectors include windows vertically mounted in walls, skylights installed on a roof, window box collectors and attached solar greenhouses. The greatest heat loss through windows occurs at night through north facing windows. Since they never receive direct sunlight, they can only lose energy. Windows should have at least two layers of glazing. Triple glazing absorbs and reflects more incoming solar energy than it saves in reduced heat loss. Double glazing should be used on south-facing windows and triple glazing elsewhere.

Solar energy can also be collected by south-facing skylights. Another technique is the solar attic, where a solid roof is replaced with glass or plastic glazing. Although the light is not admitted directly into the interior, the attic collects the entering heat which is transferred with a fan and duct work.

External wall and window box collectors are air-heating collectors that use metal absorber plates placed in boxes behind glass or plastic glazing. The heat they collect is vented into the interior of the building. Solar greenhouses should be located on the south side of the building.

One square foot of south-facing glazing can collect as much as 1,300 Btu of solar energy during a sunny, 30°F day. On such a day, a 1,500-square-foot, well-insulated building may require 250,000 Btu. A building with too much glazing and no heat storage may be too hot during the day and too cold at night. When no thermal mass is used, there should be about 0.15 square feet of glazing for each square foot of heated floor area. Windows facing away from true south can still be useful. An unshaded window oriented 25 degrees from true south will still receive more than 90% of the energy received by a south-facing window, while a window facing southwest or southeast will collect about 75 to 80%.

Beyond 45 degrees from true south, windows quickly lose their effectiveness as collectors. Vertical windows should be recessed slightly into the adjacent wall in order to reduce infiltration from crosswinds and to provide some summer shade. Additional provisions for shading and summer venting are usually required.

The window sash should be plastic since such sashes provide lower heat loss. If the sash is metal, it should contain a nonconducting thermal break to reduce heat loss. A system called Beadwall by Zomeworks of Albuquerque, New Mexico, uses a vacuum system to fill a double glazed window with small polystyrene balls.

The airspace may be thin (Trombe wall) or thick (solar greenhouse). During the day, the sun heats the airspace, and the hot air rises and passes through vents into the living space. At the same time, the sun also heats the storage mass, which radiates its heat into the living space after the sun leaves.

In a remote thermal storage system the collector and the storage mass can be located away from the building space. During the day, the air heated in the collector passes into the space and the storage mass. At night, the mass radiates and convects its heat to the living space. A hybrid system uses passive components to collect heat and active components (fans, pumps, ducts) to distribute the heat to the building space.

SOLAR HEAT STORAGE

A thermal mass may be masonry, rock, water or phase-change materials. It provides storage and reduces overheating. The thermal mass lessens temperature variations in the space by absorbing the surplus heat and releasing it when there is not direct solar energy.

Water absorbs heat in a uniform, constant manner. Temperatures at the surface change more slowly and over a narrower range than is the case with masonry storage. A narrow range of surface temperatures is desirable because a surface that is very warm will lose more heat back through the glazing. A cubic foot of water weighs about half as much as the same volume of bricks but its greater heat capacity can store three times as much heat. Masonry is often used for direct heating by sunlight and rock for convective heating by air. Rock storage is regularly used in remote thermal storage systems or when the storage component cannot be near the collector. Rock storage is useful for air-heating collector systems since a rock-filled bin is porous and has more surface over which to transfer heat.

Phase-change materials (PCM), such as eutectic salts and paraffin, are chemical compounds that melt between 80 and 120°F. The heat fusion of these materials is great in comparison to the heat capacity of water or masonry. Small volumes of PCMs can provide heat storage equivalent to much larger volumes of water or masonry. Phase-change materials are generally packaged in containers or incorporated into ceiling tiles.

Phase-change materials are not useful in systems where the maximum temperatures do not often exceed the material's melting point. There have also been problems with some types of materials failing to solidify after many cycles of heating and cooling. This has generally happened in large containers.

The use of phase-change materials in large containers also depends on the water volume and surface area along with the window area, space heat loss and container surface color. The wall should receive direct sunlight four or more hours daily and should contain at least 1.1 cubic feet of liquid (8 gallons) per square foot of glazing. Containers should be a dark color to maximize absorption. If transparent or translucent containers are used, a dark dye can be added to the water.

Phase-change materials can be incorporated into structural materials via ceiling tiles and wall panels. These provide a lighter weight thermal mass. A distribution system can transfer heat to other rooms in the building. This is accomplished by vents or ducts through walls and fans to move the air. Supply and return vents and ducts allow the air to be recirculated.

INDIRECT-GAIN SYSTEMS

This technique avoids the problems of glare, overheating and the fading of furnishings with a direct-gain system by placing the heat storage component next to the glazing and in front of the living space. A masonry exterior wall can be used as storage with a double layer of glazing over it. The storage wall provides radiative heating to the space. The thickness of the wall is critical since a wall that is too thick will cause lower space temperatures and heat at the wrong time. One that is too thin will cause overheating.

A thermal storage wall can be masonry storage placed 3 to 4 inches behind a glazing. There is no venting through the wall into the living space. Heating of the space is by radiation from the wall. Sunlight heats the wall directly and the heat takes some time to pass through the wall. This is the thermal lag. If the wall has the proper thickness, heat will radiate from the interior side at night. The space is heated only slightly during the day. Daytime heating is provided mainly by heat collected during the previous day.

DIRECT-GAIN SYSTEMS

In a direct-gain system the sunlight passes through the glazing and heats objects and people in the space and falls upon the storage component. With large areas of glazing these systems can easily overheat rooms if they do not contain sufficient thermal mass to maintain a relatively constant temperature with the living space. The thermal mass is generally part of the floor or northern wall and is positioned to receive direct or reflected sunlight for most of the day. When the space is warmer than the mass, with the mass in direct sunlight, heat will tend to flow into the mass. If the space temperature falls below that of the mass, heat radiates from the mass into the space. In this way, the mass acts like a thermal damper, reducing the temperature variations within the space.

Depending upon the climate, a direct-gain system supplying daytime heat requires about 0.1 to 0.2 square feet of glazing for each square foot of space floor area. System efficiency depends upon how well the glazing, space and storage are matched.

A large glazed area collects large quantities of heat, but if the heat storage capacity is too small, the space will overheat quickly on a sunny day. If the heat storage capacity is too large, the space temperature may not get hot enough.

About 4 cubic feet of exposed masonry, placed along the back or side wall of the room or as an exposed floor, should be used for each square foot of glazing that admits direct sunlight. This quantity of masonry will limit indoor temperature fluctuations to about 15°F. Larger amounts result in smaller fluctuations. A wall color of flat black is best for heat absorption but dark blue or red is also useful.

If a water wall is used, the temperature fluctuations will depend upon the water volume and surface area along with the window area, space heat loss and container surface color. The wall should receive direct sunlight four or more hours daily and should contain at least 1.1 cubic feet of liquid (8 gallons) per square foot of glazing. Containers should be a dark color to maximize absorption. If transparent or translucent containers are used, a dark dye can be added to the water.

Phase-change materials can be incorporated into structural materials via ceiling tiles and wall panels. These provide a lighter weight thermal mass. A distribution system can transfer heat to other rooms in the building. This is accomplished by vents or ducts through walls and fans to move the air. Supply and return vents and ducts allow the air to be recirculated.

THROMBE WALLS

The Trombe wall was named after the French inventor Felix Trombe in the late 1950s. A Trombe wall is similar to a nonvented thermal storage wall except that it has vents at the floor and ceiling to allow the space to be heated both day and night. During the day, the vents are left open. Heated air rises through the air space between the glass and the wall and passes into the space through the upper vent. Cooler air from the space is returned to the air space through the lower vent. At the same time the wall absorbs solar energy and heat moves by conduction to the interior face of the wall.

At night, the vents are closed, and radiative heating from the wall warms the space. To ensure that heated air moves only in the direction of the living space, anti-backdraft dampers may be used in the vents.

A typical Trombe wall consists of an 8- to 16-inch-thick masonry wall coated with a dark, heat-absorbing material and faced with a single or double layer of glass. The glass is placed from 1 to 2 inches from the masonry wall to create a small airspace. Heat from sunlight passing through the glass is absorbed by the dark surface, stored in the wall, and conducted slowly inward through the masonry.

Applying a selective surface to a Trombe wall improves its performance by reducing the amount of infrared energy radiated back through the glass. The selective surface consists of a sheet of metal foil glued to the outside surface of the wall. It absorbs almost all the radiation in the visible portion of the solar spectrum and emits very little in the infrared range. High absorbency turns the light into heat at the wall's surface, and low emittance prevents the heat from radiating back towards the glass. Although not as effective as a selective surface, painting the wall with black, absorptive paint can also help the wall to absorb the sun's heat.

For an 8-inch-thick Trombe wall, heat will take about 8 to 10 hours to reach the interior of the building. Heat travels through a concrete wall at a rate of about one inch per hour. This means that rooms remain comfortable through the day and receive slow, even heating for many hours after the sun sets, greatly reducing the need for conventional heating. Rooms heated by a Trombe wall often feel more comfortable than those heated by forced-air furnaces because of the radiantly warm surface of the wall, even at lower air temperatures.

Trombe walls can be used in conjunction with windows, eaves, and other building design elements to evenly balance solar heat delivery. Strategically placed windows allow the sun's heat and light to enter a building during the day to help heat the building with direct solar gains. At the same time, the Trombe wall absorbs and stores heat for evening use. Properly sized overhangs shade the Trombe wall during the summer when the sun is high in the sky. Shading the Trombe wall prevents the wall from getting hot during the time of the year when heating is not needed.

Commercial and residential buildings with Trombe walls include the Solar Energy Research Facility, Zion National Park Visitor Center, Grand Canyon house and Van Geet house. These walls were designed using computer software such as SERI-RES or BuilderGuide, which is available through the Passive Solar Industries Council.

At the Solar Energy Research Facility, a Trombe wall warms the shipping and receiving area where conventional heating methods would have been expensive. The 20 by 35 foot wall uses concrete to store and distribute heat. During the summer, when the sun is higher in the sky, most of the sunlight reflects off the glazing to keep the building's interior relatively cool. The Trombe wall is one of several passive solar features that reduce building energy use by 30% compared with conventional construction.

Another type of Trombe wall has a zigzag design to reduce glare and excessive heat gains during the day. This undulating Trombe wall has three sections. One of the sections faces south and the other two are angled inward in a V shape. On one side of the V is a southeast-facing window that provides both light and direct heat gain in the morning when quick heating is most needed. On the other side of the V is a Trombe wall that stores hot afternoon sun for redistribution in the evening hours. The wall also uses exterior overhangs to shade it from the hot sun during summer.

In Grand Canyon National Park a foot-thick concrete wall absorbs and stores the heat from sunshine. This Trombe wall is an extension of the building's foundation, which rises three feet higher than normal to create thermal mass. The concrete is faced with a frosted double glazing. A selective surface of black metal foil reduces heat loss from radiation back through the glass. The inside surface temperature of the wall reaches 100°F by late afternoon. Heat radiates from the wall to the interior for many hours after sunset, with peak indoor temperatures occurring at about 10 p.m.

The size and thickness of a Trombe wall depends upon latitude, climate and heat loss. A moderately cold climate of 6,000 degree-days will require about 0.6 square feet of double-glazed, south-facing wall for each square foot of floor area. This can supply about 65% of the seasonal space heating requirements. Movable insulation and light-colored or reflecting surfaces can reduce the required area by about 20%.

Since an 8- to 12-inch-thick concrete wall has a thermal lag of 6 to 10 hours, at least one square foot of venting is needed at the top and one square foot at the bottom for each 100 square feet of wall area. The exterior of the wall should be a dark color for high absorptivity. Paint specifications can be supplied by the manufacturer. Movable insulation can also be added to the interior side of the wall.

WATER WALLS

Water in containers can be used instead of masonry in an indirect gain system. For the same wall size and storage capacity, water is somewhat more efficient than masonry since its surface temperature will not go as high and less heat is lost back through the glazing. Nighttime surface temperatures in a water wall are higher than in a masonry wall leading to greater heat loss.

Vents at the floor and ceiling allow the space to be heated both day and night. During the day, the vents are left open. Heated air rises through the air space between the glass and the wall and passes into the space through the upper vent. Cooler air from the space is returned to the air space through the lower vent. At the same time the wall absorbs solar energy and heat moves by conduction to the interior face of the wall. At night, the vents are reclosed, and radiative heating from the wall warms the space. To ensure that heated air moves only in the direction of the living space, anti-backdraft dampers may be used in the vents.

AIR-HEATING COLLECTORS

Air-heating collector systems use either natural or forced convection to move air through the collector and into the space. A flat plate collector made up of glazing over a dark absorber is used. Storage is usually a rock-filled bin near the space. Rock storage usually requires fan powered forced convection. During the day, air in the collector is heated by the sun and is blown into the space or the storage when the space does not require heat. The cooler air from the space or the storage is pulled back into the collector. At night, the space can be heated by blowing air through the storage and extracting its heat. The different air handling modes are thermostatically controlled by dampers that are placed in the ducts connecting the collectors, the space and the storage.

WALL AND WINDOW BOX COLLECTORS

Modular wall collectors use glass or plastic over a black metal absorber. As the sun shines on the collector, the air behind the absorber is heated. It rises and moves through vents at the top of the collector and into the building. Cooler air from the living space flows through the vents at the bottom into the collector to complete a natural convection loop. At night, the vents are closed to prevent any reverse convective airflow. Window box collectors are similar to wall collectors but are mounted in windows. They can be portable, but are limited in the available collector area since the windows must have the proper orientation. When wall collectors supply more than 30% of a building's heating load, thermal storage may be required to avoid overheating.

ATTACHED SOLAR GREENHOUSES

The heat collected by a greenhouse or sunspace can supply a significant part of a building's heating load and the enclosed space can also be used as a garden patio area. On a clear, cold day, a greenhouse collects much more heat than it loses. Traditional glass greenhouses lose large quantities of heat through walls that do not face south. When the greenhouse is placed along a south-facing wall of a building, heat losses through the north wall are greatly reduced, and the greenhouse becomes a heat source.

The structure must be thoroughly sealed and double-glazed. It should be fitted with movable insulation for use at night. The greenhouse should also have roof or gable vents to prevent overheating in the summer and vents, windows or doors into the living space.

If no heat storage is used, the greenhouse should have about 0.3 square feet of south-facing, double-glazed surface for each square foot of floor space in the adjacent spaces. If movable insulation is used at night, the greenhouse will not drop below freezing. If the greenhouse is attached to a masonry building, the common wall can be used to provide heat storage for both the greenhouse and the space. Heat storage can also be provided by containers or drums of water or a rock bin.

If the greenhouse is two stories tall, a convection loop can be set up from the greenhouse into the upper story and back out of the lower one. With the addition of ducting and blowers, a greenhouse can be built on the building roof, supplying heat to the building and a storage bin.

PASSIVE SOLAR PAYBACK

Passive systems have a wide range of payback. The length of payback depends on material and labor costs and the cost of the fuel that solar energy is replacing. Simple payback assumes a seasonal heating value of 1 square foot of south glazing to save approximately 100,000 Btu. Heating oil can supply 141,000 Btu per gallon at 100% efficiency. Natural gas can supply 100,000 Btu per therm at 100% efficiency and electric power can supply 3,413 Btu per kWh at 100% efficiency.

Passive solar systems alone do not allow close control over temperature fluctuations in the space. These fluctuations can be smoothed by using back-up heat when there is not enough solar energy available to maintain comfort.

Temperature is only one parameter determining comfort. Others are humidity, drafts, activity and dress. Comfort can be maintained at temperatures lower than customary by eliminating drafts. Air circulation during the day and at night across uncovered glass, can make a room seem colder than it actually is.

The body cools itself by perspiration and a dry atmosphere causes this perspiration to evaporate rapidly resulting in a cooling effect. Humid air prevents rapid evaporation and also retains more heat than dry air. Keeping the relative humidity at a moderately high level makes a lower air temperature more comfortable.

Humidity can be increased by using a humidifier. Cold surfaces should be covered since warm bodies radiate heat to cold windows. Covering windows with movable insulation will reduce heat loss from the room and increase the comfort level of the room.

NATURAL COOLING

If supplied with the appropriate venting, some passive heating systems can help to provide cooling. Windows and skylights can be used to provide some degree of nighttime cooling. This is done by opening lower sashes on the side of the building facing the prevailing winds and upper sashes on the downwind side. Hot air will then be forced out of the building. The appropriate inlet and outlet vents can also allow a Trombe wall to provide daytime cooling.

Optimum collector tilt angle depends upon the latitude. For space heating only, collectors should be mounted at an angle from the horizontal equal to the latitude plus 5 to 15°. For systems that also supply domestic hot water, the mounting angle should equal the latitude.

Airheating collector systems must be airtight to ensure the most efficient operation. Glazing for active solar systems may be glass, fiber reinforced plastic (FRP) and thin plastic film. Glass has a very low thermal expansion when heated. Tempered glass is more resistant to breakage. FRP glazing has moderate to high thermal expansion, which can be a problem if the glazing is not adequately supported. Corrugated FRP is stronger longitudinally and less prone to sagging. Thin plastic film (Teflon) is generally used as an inner glazing when the collector is double glazed.

Collectors can be single or double glazed, depending upon the climate and collector operating temperatures. Double glazing is generally used in climates that have more than 6,000 heating degree-days. In milder climates the reduced transmission caused by the second glazing layer exceeds the increased heat gain it causes.

In space-heating solar retrofits, the solar system is usually based upon the type of heating and heat distribution systems already in place. A system with a heat pump is more compatible with solar heating.

SYSTEM COSTS

Tax credits allow you to deduct a sizable portion of the system cost. In general, the typical associated maintenance costs for an active system are 2 to 4% of the system cost per year. Back-up fuel cost is important since it determines the simple payback on the system. If it rises unexpectedly, this reduces the payback period.

Factory-built systems consist of collectors already assembled and shipped to the site to be installed. Factory-built systems usually have a warranty from the manufacturer that provides some protection against material or system failure. A site-built system may have a more limited warranty from the contractor, which may or may not cover all the components. System sizing depends on the solar collection and distribution efficiency and the heating requirements of the building. Storage sizing depends on the solar system size and the number of days of heat storage required for reserve.

COLLECTORS

The most common collector is the glass-covered flat plate type. Others include concentrating trough and tube-over-reflector collectors, and parabolic collectors. A flat plate collector uses a flat black absorber plate in a container box. Insulation behind the absorber plate and on the sides of the box reduce heat loss. The glazing may be flat glass, translucent fiberglass or clear plastic, on the sun side of the collector, 1 to 2 inches above the absorber. If the collector fluid is a liquid, runs of black pipe are attached to the metal absorber. If the fluid is air, the absorber is often textured or uses fins to increase the area that can absorb heat. The absorber may also be a blackened screen material through which the air can flow or an extruded EPDM rubber sheet with parallel water tubes running through it, which is called an absorber mat.

Concentrating collectors include tube-over-reflector types with tubes running in a north-south direction. Each tube absorber is enclosed in a glass pipe. This system can collect sunlight even when the sun is very low and is useful throughout the day and the year. Trough collectors use two mirrored sheets of glass set at right angles to make a channel.

The sheets reflect sunlight onto a blackened pipe set about two pipe diameters from the sides of the glass. The absorber can be enclosed in glass pipe, or the entire channel may be covered with a flat glass plate.

LIQUID SOLAR SYSTEMS

Liquid systems may be used for both space and water heating. In a closed-loop system the collector-to-storage loop is pressurized by a pump and isolated from the heat distribution part of the system. This type of system is always full of liquid. In climates where freezing weather occurs, it must use an antifreeze mixture. The collector fluid is passed through a double-walled heat exchanger in the hot water storage tank in order to isolate the antifreeze mixture from the hot water system.

On a pitched roof, the roof plane should not face more than about 30° from true south. Beyond this angle collector efficiency decreases significantly during the coldest months. In this case, it might be possible to tilt the collectors.

ACTIVE HOT WATER SYSTEMS

Active solar hot water systems are available with air, liquid or liquid vapor collector fluids. Liquid is the most common. The production of hot water with an air-heating collector is usually done in large space-heating systems. It requires the installation of an air-to-water heat exchanger in an air duct or inside a rock storage bed. Losses during the heat transfer process are high and air systems do not perform as well as liquid units for heating water. Air systems are generally capable only of preheating water from 70 to 95°F.

Liquid systems can use ordinary water or a water/antifreeze mixture. The closed-loop water/antifreeze package is used in colder areas where freeze protection is needed. A well designed drain-back system will automatically empty out when not collecting. Since hot water is required all year, the optimum angle from horizontal is equal to the latitude. The simplest hot water system is the water drain-back system.

In a drain-back system, the collector loop is filled by a pump. When the pump shuts off, the fluid drains into an unpressurized storage tank. Sometimes a pressurized heat exchanger collects heat from this tank and delivers it to the space.

In a drain-down system, the collector loop is pressurized by the water supply line pressure. This system drains as freezing temperatures approach. A signal from a thermostat causes a motorized valve to close off the line pressure and open a drain line. No heat exchanger is needed since potable water is circulated through the collector and distribution system.

A refrigerant fluid involves liquid and vapor circulation in an active (pumped) or passive (thermosiphon, natural convection) mode. The collector refrigerant is easily vaporized at low temperatures. It is kept at a low pressure so even when exposed to moderate temperatures, as might occur on cloudy days, the fluid will vaporize. The vaporized fluid moves through a heat exchanger in a pressurized storage tank, where it condenses and transfers its heat to the water.

In active solar systems, there are a number of factors to be considered. Solar heat may handle space heating or domestic hot water or both. The site must be suitable and the system must be compatible with climate conditions. A back-up system may need to be integrated with the solar system. Sizing the system and storage is critical. Site obstacles to solar access include trees and other buildings. Other factors are building orientation and roof pitch angle. If collectors are mounted on a flat roof, they can be rack-mounted and oriented to true south. If the system is installed on a pitched roof, the roof plane should not face more than about 30° from true south. Beyond this angle collector efficiency decreases significantly during the coldest months. In this case, it might be possible to tilt the collectors to bring them to a true south orientation.

An active solar hot water system can be backed up with electricity or natural gas. In a building with an oil-fueled heat system, electricity is the more common back-up since it is easier to install. A hot water heat pump may also be used.

Most available solar hot water systems use flatplate collectors. Systems with concentrating collectors are used if temperatures higher than about 140°F are required. The same type of hot water storage used with active space-heating systems is used with hot water systems.

Absorbers in a water system include copper or aluminum absorber plates or extruded rubber absorber mats. These mats come in long rolls and can be cut to the desired size. In some applications such as heating swimming pools, the mat is used without glazing or framing. More commonly, an absorber mat is substituted for the metal absorber plate. It is less expensive than standard absorbers.

The minimum operating temperature of an active solar system should equal the minimum operating temperature of the backup distribution system when solar and backup use the same distribution loops. Typical minimum temperatures are 95°F for air-heating systems and 140°F for water-heating systems. Most water drain-back systems, however, operate at a temperature between 85 and 120°F.

Generally, the lower the operating temperature of the system, the higher the collector efficiency. This is because heat loss from the collector decreases as the difference between the temperature of the collector and ambient temperature decreases.

Sizing of an active solar hot water system depends on solar collector efficiency and the hot water requirements of the system. Storage sizing also depends upon hot water use.

PASSIVE HOT WATER SYSTEMS

Passive systems tend to be less complicated than active systems. Two types of passive systems are used: the storage or breadbox water heater and the thermosyphoning or hydrosyphoning water heaters with solar panels.

The storage type of preheat system consists of a blackened collector-storage tank. The water is heated by the sunlight that falls on the tank. Cold water from the water line is fed into the tank at the bottom, and warmed water is then removed from the top and piped to a standard hot water tank where it is boosted to the necessary temperature.

Unless the system is insulated, it loses heat at night and does not provide much heat on cloudy days. To avoid this, the storage tank can be placed in an insulated box with a glazed top and side. This is called a breadbox water heater. During the day, the top and side doors of the box are open, allowing sunlight to fall on the tank. At night, the box is closed, keeping in the heat.

Water heated by a breadbox heater should only require significant boosting by a standard water heater during the middle of winter. During the remainder of the year, water from the breadbox should be sufficiently hot for direct use.

In the thermo- or hydrosyphoning type, an insulated storage tank is mounted at least a foot higher than several single-glazed, flat plate collectors. Cold water is piped into the bottom of the storage tank and fed from the bottom of the storage tank into the lower end of the collector panels. As it is heated in the panels, the water rises and enters the top of the storage tank. Hot water is withdrawn from the top of the storage tank. No pumps are required.

The water in the tank will remain hot all night if the weather is mild and the demand for hot water is not too great. It is also possible to mount the panels on the lower part of the roof and the storage tank inside. The thermosiphoning solar hot water system is widely used in Israel and Japan, where the collectors are placed on the roof.

In solar hot water systems, the manufacturers may use the specifications to make the system appear more effective than it really is. The amount of gallons of hot water the system will produce per day depends on how much sun is available to produce hot water.

SOLAR COOLING

The heat gained from sunlight may be used for air conditioning. This principle has been used for many years in gas refrigerators and gas air conditioning units. Using a heat source for cooling works on the principle of latent heat.

Latent heat, which means hidden or unseen heat, refers to the heat required to cause a change of state in matter. For example, it requires a certain amount of heat to raise a volume of water to the boiling point. If it takes 8 minutes to raise the small amount of water from room temperature to the 212°F boiling point, it may take much longer to raise this water one more degree to 213°F. This is because more heat is required to turn the liquid water into its gaseous state, then is required to raise its temperature while it is still in its liquid state. This extra heat is the latent heat.

An opposite effect occurs when gases condense into liquids, they give off heat. In order to complete a change of state from liquid to gas, extra heat is required and to change a gas into liquid releases extra heat.

When this effect is combined with compressing vapors into a liquid state, the operating mechanism for electric refrigeration systems is obtained. In these systems, a refrigerant is pressurized and sent through a small orifice. As it leaves the orifice, it is vaporized into a gaseous state and it absorbs the latent heat that is required for a change of state from the surrounding area. This has the effect of cooling the tubing and air around the refrigerant. An equal amount of heat is given back to the area surrounding the compressor, where the gas is turned back into a liquid.

There is no gain or loss of heat in the system. But there is a gain of heat in one area and a loss of heat in another area. The refrigeration coil can be placed where the heat loss occurs to provide cooling.

HEAT INITIATED REFRIGERATION

In a heat-initiated refrigeration system, the compressor is replaced by two tanks that contain a liquid, into which the refrigerant can dissolve. A pressure differential is maintained between the two tanks by keeping them at different temperatures. Gases expand when heated and create a higher pressure in the hotter tank. This pressure differential has the same effect as a pump that produces a pressure difference between its inlet and its outlet.

A heat-initiated refrigeration system uses a refrigerant mixture of ammonia, water, and sodium thiocyanate (a salt). The water and sodium thiocyanate are used since they readily dissolve the ammonia. At high temperatures, this solution gives up the ammonia as a vapor, without the water and salt solution vaporizing.

The solution can be heated by any effective means including solar-derived heat. Ammonia is vaporized from the warm ammonia solution. If the two ammonia solution tanks are located at different levels because of the elevation difference, the pressure of the ammonia solution in the lower tank will be higher than the pressure in the higher tank, even without the temperature difference.

The ammonia vapor moves to the condensation tank where enough heat has been dissipated to allow the ammonia to condense. The condensation gives off heat, which must be carried by the surrounding air, or by some other means. As the ammonia condenses in the condensation tank, it passes through a pressure reduction valve, and into a cold chamber where the pressure is the same as the higher ammonia solution tank. Since the pressure is lower, the condensed ammonia boils, absorbing heat from the surrounding areas. This tank becomes a cooling source.

Some adaptations of this principle use a lithium bromide solution and a U-shaped tube. The partially vaporized solution is in one leg and the liquid in the other with a condenser between them.

SOLAR CONCENTRATORS

Solar concentrators contract the solar radiation from a relatively large area onto a small area. A parabolic mirror of four feet in diameter covers an area of 4 pi, or 12.57 square feet (1.17 square meters). This surface area is measured on a plane and is somewhat less than the surface area of the curved mirror. If the sun is about 20% down from peak strength, its strength should be about 800 watts per square meter. Then the total amount of energy striking the mirror is almost a 1,000 watts.

WIND POWER

A wind machine captures energy from the wind and converts it to mechanical or electrical energy. The wind rotor may have two, three, four or more blades. The rotor turns a drive shaft connected either to a water pump or and electric generator. Electricity produced by the generator on a wind machine is sent through a cable into the electricity distribution system of a building or directly to an electric utility.

Adequate wind must be available and various legal and institutional barriers must be hurdled. Drafts may occur along the sides of the building. Tower heights may be restricted by local zoning ordinances and it may not be possible to increase the height very much. Obstructions can sometimes be helpful and create a wind tunnel, funneling winds into a narrow gap and creating an area of higher wind velocity. You may be able to use such increased wind velocities to great advantage if the effect is a fairly constant one. Such a site is likely to be highly turbulent at times and that excessive turbulence can diminish performance or even destroy a wind machine.

The cost of electricity from a large machine (10 kilowatts) will be less than from a small machine (2 kilowatts). A 1-kilowatt machine in Boston will generate about 2.7 kilowatt-hours of electricity per year, or about 30% of the maximum possible output of the machine.

In some areas, under more favorable conditions, a 10-kilowatt wind generator can produce electricity at costs competitive with utility-supplied power. As the size of the wind machine increases, the cost of the produced electricity drops. Electricity prices are likely to increase at a faster rate in the coming years as oil and gas become more expensive and as more costly new generating plants are introduced into utility rate structures. Efforts to implement energy conservation measures in order to limit the need for new generating capacity may also help increase the cost of power by reducing overall demand for electricity. In order to maintain their profitability, utilities may find it necessary to ask for larger rate increases.

Wind machines purchased in the future will be more expensive than they are today, and it will take a longer time for their generating costs to become competitive with utility-supplied electricity. As the price of electricity rises, the payback for a wind machine becomes shorter. As machine output goes up, electricity costs go down. Some wind machines maintain a fairly constant efficiency, while others may lose efficiency.

The most important factor for a wind system is the availability of wind. The higher the average wind speed, the better since the energy content of the wind is equal to the cube of the velocity. A doubling of wind velocity provides an eight-fold jump in available wind energy. An increase in average wind speed from 12 mph to 13 mph, which is only an 8% increase in wind speed, results in a 27% increase in wind energy. The relative proportion of high and low wind speeds is important. A site where the wind velocity is 8 mph half of the time and 16 mph the other half results in an average wind speed of 12 mph and about 2.5 times as much wind energy as a site with only 12-mph winds. A larger machine will usually represent a more favorable investment than a small one. State solar tax credits can also bring the costs of wind machine electricity while furling and unfurling ribbons indicate much turbulence. One way to avoid turbulence is to increase tower height. A wind machine should be raised at least 25 to 30 feet higher than any obstructions within a 300- to 500-foot radius. If the obstruction is large, it may be necessary to go even higher. It may be preferable to change the site as far upwind or downwind as possible from nearby obstructions. More turbulence can be expected if the machine is erected on a building roof since updrafts and downdrafts occur along the sides of the building. Tower heights may be restricted by local zoning ordinances and it may not be possible to increase the height very much. Obstructions can sometimes be helpful and create a wind tunnel, funneling winds into a narrow gap and creating an area of higher wind velocity.

At very high wind speeds, a wind machine must stop operating, to protect it against damage. The velocity at which this occurs is called the cut-out velocity. You may be able to use increased wind velocities to great advantage if the effect is a fairly constant one. Such a site is likely to be highly turbulent at times and that excessive turbulence can diminish performance or even destroy a wind machine.

In the past there has been some resistance by utilities to wind machine installation and connection into the power grid. There are local zoning ordinances, building codes and state regulations that might restrict or forbid wind machine installation.

PURPA

Power companies are required by the Public Utilities Regulatory Policy Act (PURPA), 1978, Sections 201 and 202 and the National Energy Conservation Policy Act, 1978 to purchase surplus power from small power producers at a fair price. Utilities must also supply back-up power. Utilities have argued that a wind energy system will reduce a customer's total demand for electricity while leaving the peak demand unchanged. The customer will generally purchase less electricity from the utility but, when the wind is not blowing, may still require the generating capacity to meet peak demand, which may or may not occur during the utility's peak demand period. If you wish to sell excess power to a utility, you may need to install a synchronous inverter. You may also down and can shorten the payback time.

The two basic types of wind machines are horizontal axis and tical axis. Horizontal axis machines are similar to the traditional, fa windmills used to pump water. In vertical axis machines, the drive is vertical. Horizontal axis machines are usually self-starting, while vertical axis machines are not. Vertical axis machines will operat gardless of wind direction, while horizontal axis machines must b tated into the direction of the wind. The vertical axis allows the gene to be placed at ground level, while the horizontal axis design us requires the generator to be mounted on a tower. Most commerc available wind machines are horizontal axis.

The ability of a wind machine to produce energy from the depends upon rotor area, blade design and conversion efficiency. area swept by the rotor determines the amount of energy that ca produced. Since the area of a circle increases as the square of the ra the amount of energy that can be extracted from the wind increases a square of the blade length. If the blade length is 5 feet, the blade at $\sim 5 \times 5 \times 3.14 = 75$ square feet.

The efficiency of a wind machine is the fraction of available energy converted into electricity. This depends on the blade, transmi between the rotor and the generator and the design of the generatoi ideal wind machine may extract about 2/3 of the energy available i wind. Practical wind machines do not attain efficiencies this high : some energy is lost due to friction. Most horizontal axis machines not begin turning until the wind reaches a velocity of about 7 to 10 Wind machines do not reach their maximum power and efficiency the wind is blowing at the rated velocity, which is generally abou mph.

Turbulence is uneven wind flow caused by obstructions suc hills, trees and buildings. Too much turbulence exposes a wind mac to rapidly changing, even dangerous, stresses that can damage it. degree of turbulence at the site can be checked with an ordinary Remove the tail and fasten a string with ribbons at least 3 feet attached at 10-foot intervals. When the kite is in the air, the end oi ribboned string is held firmly on the ground in a vertical position. behavior of the ribbons in the wind will indicate the degree and he of turbulence. Smoothly streaming ribbons indicate little turbul be required to pay a charge for a second electricity meter to monitor the amount of power fed by the machine into the grid.

Some wind generators produce direct current (DC), rather than the 60-cycle alternating current (AC) which is used in the United States. Figure 2-1 shows the main elements of a 12-volt DC wind and photovoltaic system.

For AC loads, it is necessary to convert DC to AC, which is done with an inverter. One conversion device is the synchronous inverter, which may be line or self-commutated. It is triggered by the utility and synchronizes its output with the utility power. A synchronous inverter will not be required if the machine has an AC induction generator with a 60-cycle, 120-volt AC output.

Wind machines equipped with induction generators are controlled by the electricity from the utility and thus require no inverter. Such systems may be more cost-effective and less complex than a DC wind machine connected to an inverter.

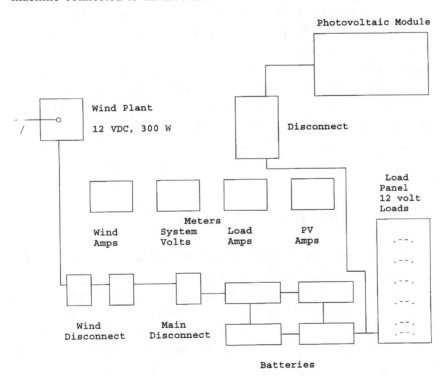

Figure 2-1. 12-volt DC wind and photovoltaic system

SYNCHRONOUS GENERATORS

The synchronous generator is commonly used for mid-sized independent generating systems. These generators have a rotating field and are also commonly known as alternators. The armature windings, which may be at the center of the generator, are stationary and the field windings are rotated. The electricity that is produced by these units is taken directly from the stationary armature and no slip-rings or brushes are required.

In a synchronous generator the frequency is determined by the speed of the field's rotation while the voltage is determined by the amount of current flowing through the field windings. The amount of current produced depends on the amount of torque required to move the field windings. When the load requires more current, the generator shaft will become more difficult to turn. When the load needs less current, the field windings become easier to turn.

The current that flows through the field windings is direct current and is used to setup the magnetic field in these windings. The synchronous generator produces alternating current when the magnetic fields produced by the DC currents in the field windings pass through the armature windings. The magnetic field causes a current in one direction and then in the other direction. The output voltage is controlled by the excitation in the field windings. A voltage regulator checks the output voltage and compares it to a standard reference voltage. If the load varies, the excitation is adjusted to keep the voltage constant.

SOLAR THERMAL GENERATION

If the sun's rays are concentrated using mirrors, high enough temperatures can be generated to boil water to drive steam engines. These can produce mechanical work for water pumping or more commonly for driving an electric generator. The systems used have a long history with the first prototypes built over 100 years ago. If cheap oil and gas had not appeared in the 1920s, solar engines might have developed to be more common.

The first solar engines appeared in France in the 1870s and 1880s in a series of machines ranging from a solar printing press to solar wine stills, solar cookers and solar engines driving refrigerators. Their basic collector design was a parabolic concentrator with a steam boiler mounted at the focus. Steam pipes ran down to a reciprocating engine. These systems were widely acclaimed but they suffered from the low power density of solar radiation and a low overall efficiency. One machine occupied 40 m of land to drive a one-half horsepower engine. By the 1890s, it was clear that this was not going to compete with new supplies of coal, which were appearing as a result of increased investments in mines and railways.

At the beginning of the 20th century, a U.S. inventer named Frank Shuman built a large system at Meadi in Egypt. It used five parabolic trough collectors, each 80 m long and 4 m wide. At the focus, a finned cast iron pipe produced steam for an engine. By 1913, Shuman wanted to build 20,000 square miles of collectors in the Sahara, which would produce the 270 million horsepower required to equal all the fuel mined in 1909. Then along came World War I followed by the era of cheap oil and interest in solar steam engines died.

The steam engine is a vapor cycle engine, which works by boiling a fluid to produce a high-pressure vapor. This goes to an expander, which extracts energy and emits low-pressure vapor. The expander may be a reciprocating engine or a turbine. Figure 2-2 shows a vapor cycle turbine.

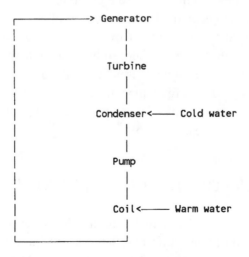

Figure 2-2. Vapor cycle turbine

In order to boil water, its temperature must be raised to at least 100°C. This may be difficult to achieve with simple solar collectors. It is more appropriate to work with a fluid with a lower boiling point. This requires a closed cycle system with a condenser that turns the exhaust vapor back to a liquid and allows it to be returned to the boiler.

All heat engines produce work by taking in heat at a high temperature, T_H, and ejecting it at a lower one, T_L

The maximum ideal efficiency is $1 - T_L/T_H$

The temperature must be expressed in degrees Kelvin (degrees Centigrade plus 273). A turbine fed from parabolic trough collectors might take steam at 350°C and eject heat to cooling towers at 30°C. Its theoretical efficiency would be $1- (30 + 273)/(350 + 273)$, or 51%. Its practical efficiency is more likely to be about 25%. The theoretical efficiency of a heat engine that was fed with relatively low temperature vapor at 85°C and exhausted heat at 35°C is only 14%.

POWER TOWERS

In the early 1980s, large, experimental electricity generation systems were built that make use of high temperatures. Several used a central receiver system or power tower. The 10-megawatt (MW) Solar One system at Barstow, California, used a group of tracking heliostats, which reflect the sun's rays onto a boiler at the top of a central tower. To carry away the heat at temperatures of over 500°C, these systems use either special high-temperature synthetic oils or molten rock salt. The heat transfer medium needs a high thermal capacity and conductivity. The hot salt is used to produce high-temperature steam to drive a turbine. Other large experimental systems of this type have been built in Sicily, Spain, France and the Crimea.

PARABOLIC TROUGH CONCENTRATOR SYSTEMS

Solar-generated electricity has been produced at solar power stations in the Mojave desert in California. Between 1984 and 1990, Luz International built nine systems with 13 to 80-MW rating. These are massive versions of Shuman's 1913 design, using large fields of parabolic trough collectors. Each successive project focused on economies of scale.

The 80-MW system used 464,000 square meters of collector glass. The collectors heat synthetic oil to 30°C, which then produces high-temperature steam using a heat exchanger. This temperature gives a solar-to-electricity conversion efficiency of 22% peak and 14% on average over the day. This is close to commercial photovoltaic systems.

The economies of scale mean that the system could use well-developed, megawatt steam turbines normally used in conventional power stations. Estimated generation dropped from 28 cents per kWh for the first Luz system to 9 cents per kWh for the eighth unit. The Luz systems were intended to meet peak afternoon air conditioning demands in California.

PARABOLIC COLLECTORS

A direct beam intensity of 800 W/square meter is roughly equal to the energy density of the boiling ring on an electric cooker. The common method of concentrating the solar energy is to use a parabolic mirror. All rays of light that enter parallel to the axis of a mirror formed in this shape will be reflected to one point, the focus.

A variant is the Winston mirror which concentrates the solar radiation to a more inexact focus, but this is at the back of the mirror, which is flat instead of curved. These mirrors can be made for line focus or point focus. The line focus type is also called a trough collector since the energy is concentrated on a small region running along the length of the mirror. The collector normally faces south and needs only to track the sun in elevation.

The point focus type concentrates the energy in the center of the mirror. For optimum performance, the axis must be pointed directly at the sun at all times, so it needs to track the sun both in elevation and in azimuth. A parabolic collector can achieve a concentration ratio of over 1000 while a line focused parabolic trough collector may achieve a concentration ratio of 50, which is still adequate for a power system. A line focused parabolic trough collector can produce a temperature of 200-400°C while a dish system can produce a temperature of over 1500°C.

PARABOLIC DISH CONCENTRATOR SYSTEMS

Instead of moving solar heat from the collector to a separate engine, an alternative approach is to place the engine at the focus of a mirror. This has been done with small steam engines and with Stirling engines. Stirling engines were invented in 1816 and competed with steam engines in the latter half of the 19th century. They have few moving parts and use a piston where the heat is applied to the outside of the cylinder. This heat expands the working gas, which can be air, nitrogen or helium, and this moves the main power piston.

A second piston, the displacer, shifts the gas to the cold side of the engine, where it contracts, and moves the power piston back again. Steam engines are operated with input temperatures below 700°C, but Stirling engines, built with the right materials, can be made to operate at temperatures of up to 1000°C, with accompanying higher efficiencies. Experimental solar systems have achieved overall conversion efficiencies, approaching 30% on average. Stirling engines can also run on biofuels. Since combustion can take place outside the engine, it is easier to optimize combustion conditions, and the engines can tolerate a wider range of fuels than internal combustion engines.

The parabolic concentrating mirrors do not have to be made of glass. Circular sheets of aluminized plastic film can be bent into a parabolic shape. This produces a lightweight mirror, which requires a lightweight structure to support and track it.

In active solar systems, there are a number of factors to be considered. Solar heat may handle space heating or domestic hot water or both. The site must be suitable and the system must be compatible with climate conditions. A back-up system may need to be integrated with the solar system. Sizing the system and storage is critical.

SOLAR SITES

Obstacles to solar access include trees and other buildings. Other factors are building orientation and roof pitch angle. If collectors are mounted on a flat roof, they can be rack-mounted and oriented to true south. If the system is installed on a pitched roof, the roof plane should not face more than about 30° from true south. Beyond this angle collector efficiency decreases significantly during the coldest months. In this case, it might be possible to tilt the collectors.

SOLAR PONDS

Another approach to solar thermal electricity production is the solar pond. Even when a layer of ice floats on the surface of a solar pond, the water at the bottom is still near 80°F. By the first day of summer it will be warm enough to heat an outdoor swimming pool and by early October, the pond can still be used for another 3 months to heat a building. The cost for this energy is less than the cost of burning heating oil at 75% furnace efficiency.

Solar ponds are now operating in Israel and a few U.S. sites. Saline ponds are now producing electricity in New Mexico, Nevada, and Virginia. They are quick and inexpensive to build compared to other energy sources. Solar ponds can supply heat, electricity, or both with low-maintenance.

Large, naturally-occurring brine ponds with salt gradients provide a model for the solar pond. Typically, solar ponds are one to three meters deep, and contain water in layers of differing densities. The top layer is relatively fresh, and is the least dense. The thin middle layer increases in salinity and density with depth. The bottom layer is the heaviest and saltiest. As solar radiation penetrates the water, the light is converted to heat, and the heat is trapped in the bottom, saline layer.

Two factors prevent the heat from rising and escaping. It is held by the salt in the densest layer, which limits its movement. In the middle the boundary layer acts as a density and salinity gradient and becomes an insulating blanket. The pond is both solar collector and storage medium.

The saline gradient is the critical layer of the pond, since it is the only one in which convection does not occur. The top layer is exposed to air and wind and is affected by changes in atmospheric ambient temperature and surface movements. Some of its heat is lost through convention. The bottom layer loses a small amount of energy to the ground by conduction, creating a temperature difference, which generates a small convection pattern. The middle layer is non-convecting and prevents heat loss through bottom layer convection.

The saline gradient is formed by placing a layer of fresh water on top. The layer boundaries can be maintained by injecting concentrated brine at the pond bottom, and fresh water at the top.

One way to tap the heat stored in the saline pond is by running a heat exchanger through the water at the bottom of the pond. The extracted heat can then be used to condition space or water.

A 2,000-square-meter pond will require about 1,100 tons of salt. From October to December, this size of pond can provide about 800 million Btu for heating. Ponds can also run absorption chillers during summer months.

Another way to tap a solar pond's heat is by piping the saline layer through an external heat exchanger, rather than one in the bottom layer. This method requires some attention to avoid disrupting the gradients. But, the heated water could be used in organic Rankine cycle engines to generate electricity. Ponds can produce about one MW (peak) of electricity per 50 acres of pond area.

The Isralis began operating a 150-kW power station in 1979 at Ein Bokek on the Dead Sea. The plant collects heat in a rubber-lined 70,000 foot^2 (6,500 m^2) pond, and sends the heated saline water through a heat exchanger, where the heat is transferred to a high-pressure organic fluid. The heated fluid expands through a Rankine cycle generator to produce electricity. Another heat exchanger in the pond's upper layer cools the organic fluid for recirculation.

Ponds producing electricity have a continuous operating capacity that is far below peak output. The pond at Ein Bokek is rated at 150-kW, but can sustain only 35-kW in summer at continuous operation, and drops to 15-kW in winter.

Solar ponds present some environmental hazards. As the sand used in the foundation settles, the pond liner is strained and may leak. Agricultural lands or local water supplies near ponds could be contaminated.

OCEAN THERMAL ENERGY CONVERSION

Ocean thermal energy conversion uses the sea as a solar collector. It exploits the small temperature difference between the surface and the colder water at the bottom. In deep waters of 1000 m or more, this can reach 20°C. Recently, large-scale experiments have been carried out in the Pacific with some success. An OTEC station producing 100-MW of electricity needs to pump almost 500 cubic meters per second of both warm and cold water through the heat exchangers while moored in sea 1000 meters deep.

PHOTOVOLTAIC SYSTEMS

The total annual solar energy input to the earth is more than 15,000 times as great as the earth's current yearly use of fossil and nuclear fuels. The term photovoltaic comes from the Greek word for light, photos and voltaic which refers to the unit of electromotive force. The discovery of the photovoltaic effect goes back to the French physicist Becquerel, who conducted experiments with a wet cell battery and found that the battery voltage increased when its silver plates were exposed to sunlight.

The first report of the PV effect in a solid substance was made in 1877 when two Cambridge scientists observed the electrical properties of selenium when exposed to light. In 1883 a New York electrician, built a selenium solar cell that was similar to the silicon solar cells of today. It consisted of thin wafers of selenium covered with thin, semi-transparent gold wires and a protective sheet of glass. Less than 1% of the solar energy reaching these early cells was converted to electricity. But, these selenium cells came into widespread use in photographic exposure meters. In the 1950s high-efficiency solar cells were developed with semiconductors. These non-metallic materials, such as germanium and silicon, have electrical characteristics between those of conductors and insulators. Transistors are also made from semiconductors in pure crystalline form, into which small quantities of impurities, such as boron or phosphorous, are diffused. This process, known as doping, alters the electrical behavior of the semiconductor in a desirable manner for transistors.

In 1953 a team at Bell Labs produced doped silicon slices that were much more efficient than earlier devices in producing electricity from light. They increased the conversion efficiency of their silicon solar cells to 6%. In 1958, solar cells were used to power a radio transmitter in the second U.S. space satellite, Vanguard I. Rapid progress has been made in increasing the efficiency of PV cells, and reducing their cost and weight. PV cells are widely used in consumer products such as watches and calculators and a number of PV power stations connected to utility grids are now in operation in the U.S., Germany, Italy, Switzerland and Japan. Domestic, commercial and industrial buildings now use PV arrays to provide some proportion of their energy needs.

PHOTOVOLTAIC GENERATION

Photovoltaic or solar cells generate electricity directly from sunlight using the properties of semiconductor material. A solar cell has two layers of specially treated silicon. When sunlight strikes the silicon, negatively charged electrons are removed from their orbits around atoms and move through the semiconductor material or an external electric circuit, if one is provided. This flow of electrons provides an electric current.

SEMICONDUCTORS AND DOPING

These PV cells consists of a junction between two thin layers of dissimilar semiconducting materials, known P (positive)-type semiconductor and N (negative)-type semiconductor. These semiconductors are usually made from silicon although PV cells can be made from other materials. N-type semiconductors are made from crystalline silicon that has been doped with an impurity like phosphorus. The doping allows the material to have a surplus of free electrons and negative charge. Silicon doped in this way is known as an N (negative)-type semiconductor. P-type semiconductors are made from crystalline silicon, but are doped with a different impurity (usually boron) which causes the material to have a deficit of free electrons. These missing electrons are called holes. Since the absence of a negatively charged electron is equivalent to a positively charged particle, silicon doped in this way is known as a P (positive)-type semiconductor.

P-N Junctions

A P-N junction is created by joining these dissimilar semiconductors. This produces an electric field in the region of the junction. The P-

N junction is not an abrupt change, but the characteristics change from P to N gradually across the junction.

PV Effect

When light falls on the p-n junction of a solar cell, the particles of energy, called photons transfer some of their energy to some of the electrons in the materials, moving them to a higher energy level. In their excited state, the electrons become free and produce an electric current by moving through the material. As the electrons move, they leave behind holes in the material, which can also move and create a current.

If there is an external circuit for the current to flow through, the moving electrons will flow out of the semiconductor through metallic contacts. The holes will flow in the opposite direction through the material until they reach another metallic contact. The PV cell generates a voltage from the internal electric field set up at the p-n junction. A single silicon PV cell produces about 0.5-V at a current of up to around 2.5 amperes. This is a peak power of up to about 1.25-W. The junction characteristics change from P to N gradually across the junction.

Solar cells are usually manufactured as discs about three inches in diameter and are often used in flat arrays (Table 2-1). A typical array consists of many cells mounted on a metal or plastic backing, wired together and covered with a protective glass or clear rubber coating. These sheets are rack or roof-mounted at the desired angle, connected to a power conditioner or converter and wired into the building's electric distribution system. Figure 2-3 shows a typical utility intertied photovoltaic system.

Since solar cells generate electricity on a year-round basis, the optimum angle for mounting them is equal to your latitude. A typical silicon solar cell generates a maximum of about one watt of power at noon.

About 500 square feet of solar cell arrays are needed to produce 5 kilowatts (5,000 watts) of power at noon on a sunny day. Such an array will generate about 600 kilowatt-hours (kWh) of electricity per month. This is about equal to the monthly consumption of a single-family home.

Solar cells and active collectors can be combined, with the cells acting as an absorber surface for an air or liquid-heating system. This is done in some commercial equipment where concentrators are used to

Table 2-1. Photovoltaic Manufacturing Processes

Single-crystalline Silicon	Thin-Film Copper Indium Diselenide

Ingot

Single-crystalline Silicon	Thin-Film Copper Indium Diselenide

Polysilicon preparation
  Crystal growing
  Ingot shaping
	Not required

Wafer/Cell

Single-crystalline Silicon	Thin-Film Copper Indium Diselenide
Ingot sizing Mounting Wire saw cutting Cleaning Chemical etching Phosphorous diffusion Post diffusion etch Oxidation Plasma etch Anti-reflective coating Front/back print Cell test Packaging	Cut glass Wash/deposit Mo electrode Pattern 1: Isolation wash/deposit CIG metals Selenize Chemical deposit CdS Pattern 2: Transparent conductors Pattern 3: Isolation test

Module

Single-crystalline Silicon	Thin-Film Copper Indium Diselenide
Stringing Circuit assembly Prelamination lay-up Edge trim and inspection Framing IV measurement and labeling Packaging	Attach leads Prelamination lay-up Lamination and cure framing Edge trim and inspection IV measurement and labeling Packaging

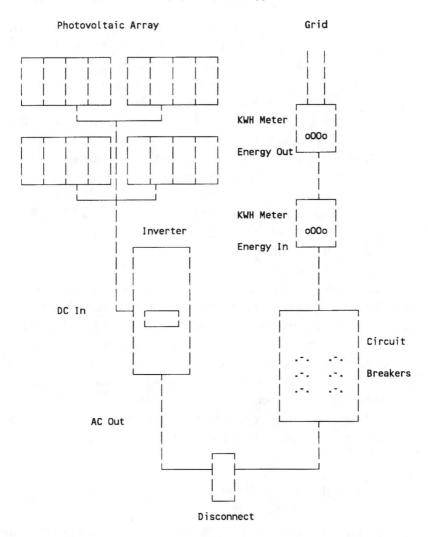

Figure 2-3. Utility intertied photovoltaic system

focus sunlight on solar cells. This reduces the collector area, and the heat produced by the cells is used for space-heating purposes. Some flat-plate combination solar cell/active heating systems are also available.

The efficiency of silicon solar cells is almost 25% in the laboratory and PV modules are available with an overall efficiency of about 18%. It is expected that modules will soon be available with efficiencies of 20% or more.

The maximum possible efficiency of a silicon solar cell in ordinary sunlight is about 27%. Until recently, this was considered the maximum efficiency but it appears possible that with the use of new materials, special arrangements of cells, filters and simple concentrators, solar cell efficiencies of 40% or more may be achieved in the future.

A simple way of viewing a typical 100-square-centimeter silicon PV cell is as a solar powered battery, one that produces a voltage of around 0.5-V and delivers a current proportional to the sunlight intensity, up to a maximum of about 2.5-3 amperes in full sunlight.

Solar cells generate direct current (DC) rather than alternating current (AC). Electronic circuitry is used to convert solar cell DC to AC. A solid-state electronic converter or inverter is used. It is possible to sell surplus electricity to your utility and only use utility power when the solar cell output is insufficient.

In the future, the costs of solar cells should be reduced through the use of new materials and manufacturing techniques under development. At the present time, solar cells are individually sliced from large cylinders of silicon. New processes, in which the cells are cut from large flat sheets, are likely to greatly reduce costs and waste. In some monocrystalline PV modules, the circular silicon slices are trimmed into squares, to increase the area of active PV material that can be included in a module. New semiconductor materials and different ways of using these materials may also reduce solar cell costs.

CELL PRODUCTION

Until recently, most solar cells were made from pure monocrystalline silicon (Si). This is silicon with a single, continuous crystal lattice structure and virtually no defects or impurities. Mono-crystalline silicon is usually grown from a small seed crystal that is slowly pulled out of a molten mass of the polycrystalline silicon using the Czochralski process, developed for transistors.

Some monocrystalline PV modules have an efficiency of about 16% and use the laser-grooved buried-grid cell technology. These cells use a pyramid-shaped texture on the top surface to increase the amount of light trapped.

Monocrystalline cells manufactured by the Czochralski process can be made from a less pure, solar-grade silicon, with only a small reduction in conversions efficiency. Solar-grade silicon can be manufactured much more cheaply than electronic-grade silicon, using several low-cost processes. Silicon can be grown in ribbon form using polycrystalline rather than single-crystal material. A thin ribbon of monocrystalline silicon from a polycrystalline or single crystal silicon melt is used in a process known as edge-defined, film-fed growth (EFG). Other PV materials such as gallium arsenide and amorphous silicon can also be used.

Solar cell wafers can be made directly from polycrystalline silicon. Polycrystalline PV cells are easier and cheaper to manufacture but they tend to be less efficient since the light-generated charge carriers (electrons and holes) can recombine at the boundaries between the grains in the silicon. Polycrystalline PV modules are also called semi-crystalline or multicrystalline and have efficiencies of about 10%. An advantage of polycrystalline silicon cells is that they can be formed into a square shape, which virtually eliminates any inactive area between cells.

Conventional silicon solar cells are several hundred microns thick. Light trapping techniques are used in thin layers or films of silicon around 20 microns in thickness. These polycrystalline thin films result in PV cells with efficiencies as high as 15% and have a low processing cost. An array of 312 Astropower modules using this technology and delivering some 18-kW was installed in 1994 at the PVUSA test site in Davis, CA.

OTHER PV MATERIALS

Silicon is not the only material suitable for photovoltaics (PV). Another is gallium arsenide (GaAs), a compound semiconductor. GaAs has a crystal structure similar to that of silicon, but consists of alternating gallium and arsenic atoms. It is suitable for use in PV applications since it has a high light absorption coefficient and only a thin layer of material is required.

They can also operate at relatively high temperatures without the performance degradation from which silicon and many other semiconductors suffer. This means that GaAs cells are well suited for concentrating PV systems. Cells made from GaAs are more expensive than silicon cells, because the production process is not so well developed, and gallium and arsenic are not abundant materials.

GaAs cells have often been used when very high efficiency, regardless of cost, is required as in space applications. This was also the case with the Sunraycer, a photovoltaic-powered electric car, which in 1987 won the Pentax World Solar Challenge race for solar-powered vehicles when it traveled the 3000-km from Darwin to Adelaide at an average speed (in day time) of 66-km per hour.

In the 1990 race the winning car used monocrystalline silicon cells of the advanced, laser-grooved buried-grid type. The 1993 winner was powered by 20% efficient monocrystalline silicon PV cells, which achieved an average speed of 85-km per hour over the 300-km course.

PV CHARACTERISTICS

When PV cells are delivering power to electrical loads in real-world conditions, the intensity of solar radiation often varies over time. Many PV systems use a maximum power point circuit that automatically varies the load seen by the PV cell in such a way that it is always operating around the maximum power point and so delivering maximum power to the load.

A typical 100 cm^2 silicon PV cell produces a maximum current of just under 3 amps at a voltage of around 0.5 volts. Since many PV applications involve charging lead-acid batteries, which have a typical nominal voltage of 12 volts, PV modules often consist of around 36 individual cells wired in series to ensure that the voltage is usually above 13-V, sufficient to charge a 12-V battery even on fairly overcast days.

In a typical monocrystalline PV module, the open circuit voltage is 21-V and the short circuit current is about 5-A. The peak power output of the module is 73-W, achieved when the module is delivering a current of some 4.3-A at a voltage of 17-V.

AMORPHOUS CELLS

Silicon can not only be formed into the monocrystalline and polycrystalline structures. It can also be made in a less structured form called amorphous silicon, in which the silicon atoms are much less ordered than in the crystalline form. A thin film of amorphous silicon is deposited on a suitable substrate such as stainless steel.

Amorphous silicon solar cells use a different form of junction between the P and the N type material. A P-I-N junction is usually formed, consisting of an extremely thin layer of P-type on top, followed by a thicker intrinsic (I) layer made of undoped material and then a thin layer of N-type amorphous silicon. The operation of the PV effect is similar to that in crystalline silicon.

Amorphous silicon cells are much cheaper to produce than crystalline silicon. It is also a better absorber of light, so thinner films can be used. The manufacturing process operates at a much lower temperature so less energy is required. It is suited to continuous production and it allows larger areas of cell to be deposited on a wide variety of both rigid and flexible substrates, including steel, glass and plastics.

The cells are less efficient than single-crystal or polycrystalline silicon units. Maximum efficiencies are around 12% and the efficiency degrades within a few months of exposures to sunlight by about 40%.

Multiple-junction devices should result in both reduced degradation and improved efficiency. Amorphous cells have been used as power sources for a variety of devices including calculators where the requirement is not so much for high efficiency as for low cost. In 1990, amorphous silicon cells accounted for around 30% of total worldwide PV sales.

Besides amorphous silicon, other thin film technologies include copper indium diselenide (CuInSe) and cadmium telluride (CdTe). Copper indium diselenide is a compound of copper, indium and selenium, which is a semiconductor. Thin film cells have attained efficiencies of about 12%.

Cadmium telluride can be made using a relatively simple and inexpensive electroplating process. Efficiencies of about 10% are possible, without the performance degradation that occurs in amorphous cells.

Another way of making PV cells is to use millimeter-sized, spheres of silicon embedded at regular intervals between thin sheets of aluminum foil. Impurities in the silicon tend to diffuse out to the surface of the sphere, where they can be ground-off as part of the manufacturing process. Relatively cheap, low-grade silicon can be used and the resulting sheets of PV material are very flexible. Module efficiencies of over 10% have been achieved.

An even more radical, photo-electrochemical, approach to producing cheap electricity from solar energy has been pioneered by researchers at the Swiss Federal Institute of Technology in Lausanne. The idea of harnessing photo-electrochemical effects to produce electricity from sunlight is extremely cheap to manufacture.

One way of improving the overall conversion efficiency of PV cells and modules is the stacked or multi-junction approach, where two (or more) PV junctions, usually of the thin film type, are layered on top of the other. Each layer extracts energy from a particular part of the spectrum of the incoming light.

CONCENTRATING PV SYSTEMS

Another way of getting more energy out of a given number of PV cells is to use mirrors or lenses to concentrate the incoming solar radiation on to the cells. Fewer cells are required depending on the concentration ratio, which can vary from as little as two to several thousand. The concentrating system must have an aperture equal to that of an equivalent flat plate array to collect the same amount of incoming energy.

The systems with the highest concentration ratios use sensors, motors and controls to allow them to track the sun in two axes, azimuth and elevation. This ensures that the cells always receive the maximum amount of solar radiation. Systems with lower concentration ratios often track the sun only on one axis and can have simpler mechanisms for orienting the array towards the sun.

Most concentrators can only utilize direct solar radiation. This is a problem in areas the solar radiation is diffused much of the time. However, some types of concentrators, such as the Winston type, allow diffused radiation as well as direct radiation, to be concentrated.

A different approach in concentrating solar energy is used in the fluorescent or luminescent concentrator. It consists of a slab of plastic containing a fluorescent dye, or two sheets with a liquid dye between them. The dye absorbs light over a wide range of wavelengths, but the light is re-radiated when it fluoresces in a much narrower band of wavelengths. Most of the re-radiated light is internally reflected from the front and back surfaces, and can only escape from the edges. Reflectors are mounted on three of the edges of the slab and on the back surface, so light can only emerge along this edge where it is absorbed by a strip of silicon PV cells. Fluorescent concentrators can concentrate diffused as well as direct sunlight. They have been used in consumer products such as clocks.

References

Harding, Jim and Elyse Axell and Others, *Tools for the Soft Path, San Francisco*, CA: Friends of the Earth, 1982.

Rosenberg, Paul, *The Alternative Energy Handbook*, Lilburn, GA: Fairmont Press, Inc., 1993.

Scientific Staff of the Massachusetts Audubon Society, *The Energy Saver's Handbook*, Emmaus, PA: Rodale Press, 1982

CHAPTER 3

INTEGRATION OF COOLING, HEATING AND POWER SYSTEMS

ENERGY AND POWER MANAGEMENT, DISTRIBUTED CONTROL TRENDS

The late 1980s and the early 1990s were characterized by slow growth and recession. Industry responded by cutting costs and drastic reorganizations. These took the form of mergers with plant closings or cutbacks, layoffs and delayed purchases for capital equipment. This reduction of personnel put pressure on the surviving departments to increase automation and become more efficient. The financial staff analyzed operations more closely and offered areas that might be improved. These economic factors along with technological advances in electronics and control hardware allowed plant automation changes that were not possible before. Among the benefits of this new technology were tools that allow users to document procedures and justify them for the next budget year.

The mission to reduce costs did not mean quality would be sacrificed. Quality needed to be improved since quality expectations and standards existed because of regulation, market competition and litigation.

DEREGULATION ISSUES

Power deregulation is a reality in several states such as California and pilot programs are going in several others. The way energy is bought is changing rapidly. Energy deregulation offers great potential for cost savings. Utility deregulation is a direct result of the Federal Policy Act of 1992.

Competitive market-based pricing is replacing state and federal rate structures. In states that are still regulated, utilities are modifying their rate structures to preserve their customer base in any future deregulated environment.

Open, competitive energy markets are unrestricted by geographical boundaries and regulated rates. The different purchasing options and rate structures are similar to what occurred following the deregulation of the telephone industry.

Advances in metering hardware, communications, and software have significantly reduced the cost of how to monitor and control energy use, even in regulated environments. Those who delay using these advances may see their costs rise significantly.

Further advances in software and communications allow facilities to link their energy costs with labor/material costs and production rates. These new tools and technologies will allow companies to negotiate better rates with utility suppliers and determine more efficient production methods and schedules. These companies will be able to find many ways to significantly lower the once fixed cost of their energy use. The new options also allow companies to guard against unexpected power reliability.

The new energy environment requires that you must know how your operations can tolerate an occasional interruption of power. Production or other operations may need to be shifted to off-peak times. Partial load interruptions or lower power quality may damage some equipment and expand maintenance costs resulting in lower productivity and lower quality.

Many industries are also facing higher quality standards. One example of higher quality are those standards enforced by the U.S. Food and Drug Administration (FDA) for pharmaceutical manufacturers. High levels of competition are also forcing improved quality. Competitive factors and improved technology mean previously acceptable levels of product quality become unacceptable. Most industrial customers require suppliers to have International Standards Organization (ISO) certification in all manufacturing processes.

Improving quality and holding down costs is proving to be a difficult challenge for those in charge of operating a building. Building efficiency requires capital investment. When money is limited, operations and maintenance functions can fall out of the bottom of capital funding requests. These cost-avoidance projects are often passed up for more lucrative, but possibly riskier, direct revenue-producing projects. Since departments must compete among themselves to receive the money they need, better documentation of operations is needed.

ENERGY MANAGEMENT

The start of networked control systems began during the energy crises of the 1970s, when the rising prices of imported oil triggered restricted energy use and more efficient energy management and control techniques. This produced the development of energy management systems (EMS) for monitoring energy usage. These systems grew over the years in both sophistication and scope.

An offshoot that appeared in the 1980s was called building automation systems (BAS). These systems added historical data, trend logging and fire and security functions to the traditional energy management functions. These applications were to focus on a return on investment based on the utility savings.

Direct digital control systems appeared in the mid-1980s and displaced older analog closed-loop schemes for temperature control. These digital systems improved both accuracy and reliability, but these systems were modeled after existing system architectures and did not include intelligent, stand-alone field devices. There were still numerous interfaces to the various building systems and the major decisions were made at a central computer.

INTEGRATED SYSTEMS

More contemporary Building Automation Systems (BAS) attempt to limit the interfaces and provide a more seamless, integrated network. Digital control networks provide an architecture that can be fully distributed with independent controllers for the systems and subsystems in a building. Ideally, all of the various components will communicate to each other in a common language.

A networked energy management system allows improvements in plug-in instrumentation, instant decision support, documentation and automation. Intelligent, versatile instrumentation at the lowest control level allows the energy management data to be easily organized for more flexible efficient management. Highly organized data collection means the facilities and operations staff can be more effective. Improved reports allow better decisions since a variety of reports can be obtained quickly.

Energy audit trails also go more quickly and can be used to demonstrate proof of performance, as well as limit future costs. A networked control system can automatically document processes, generate reports and issue work orders when problems are detected at a point in the network.

Several levels of control are generally used with several levels of hierarchy in a distributed architecture. At the lowest level are the distributed controllers. At the next or middle level is the building wide control. Next, at the highest level is information management. Each level serves its own purpose, but all levels are interconnected, similar to the operating structure of a corporation.

Distributed controllers typically employ microprocessor-based sensors and other devices to meet the needs of a specific application. These are stand-alone controllers that function as specialized tools for a specific job.

The building-wide level coordinates all of the building control strategies. This level coordinates the specialized activities and provides global direction.

In the information management level, data is collected from various points in the system and transformed into usable information. The combined efforts of the highly specialized field units are used to make information decisions about the operations of the overall system.

DISTRIBUTED ENERGY CONTROL

Distributed control means a complete control system with all the needed inputs, output and control processing logic to produce a control loop. In building control the controlled parameters include basic functions such as discharge air temperature, space temperature, humidity and fan control.

The benefits of such a control system in an intelligent, integrated heating and cooling network include repeatable and individual parameter or area (zone) control. Individual comfort control has been shown to increase employee output and provide an annual productivity gain of over $1000 per employee.

Networking takes building automation beyond traditional heating and cooling functions. Intelligent devices can be tied into the network, allowing data to be collected and energy usage to be measured. A networked system may also manage lighting, fire and access control. If these systems are fully integrated, then the expanded integrated control functions can also address environmental issues such as indoor air quality.

AIR MONITORING

Increasingly, legislation is targeted at the monitoring of volatile organic compounds (VOCs) in the atmosphere, many of which are suspected carcinogens or are acutely toxic. Monitoring requirements include rapid, multipoint, multicomponent analysis with the minimum of operator interference.

Laboratory analysis can only give a snapshot of pollutants at the time of analysis. It does not analyze exposure to pollutants during a normal working period, nor can it detect sudden chemical leaks.

Thermal desorption involves the adsorption of VOCs on materials over a certain time period followed by desorption and analysis by a gas chromatograph. Thermal desorption only reports the average concentration recorded over several hours. It does not detect short-term exposure to high levels of VOCs.

Continuous multipoint plant monitoring gives continuous information on the status of ambient air pollution at numerous locations in the plant. It provides information on levels of pollutants that workers normally are exposed to and has the ability to detect a chemical leak.

In recent years, process FTIR (Fourier transform infrared) and process mass spectrometry have been increasingly used for ambient air monitoring. Process FTIR is a powerful method for monitoring nitrogen oxides and carbon monoxide. For VOCs, FTIR only offers sensitivity at the low ppm level. Also, due to interferences from carbon dioxide and water vapor, the sensitivity to compounds such as benzene and dichloromethane is poor.

Mass spectrometers measure the masses of positively charged ions striking a detector. This allows quantification of the sample by comparison with standard calibration gases for multicomponent mixtures. Mass spectrometers can analyze a sample point in less than 10 seconds.

Mass spectrometers have proved particularly successful in Vinyl Chloride Monomer (VCM) monitoring. Other gases and sources are shown in Table 3-1. In the mid-1970s legislation was introduced in many parts of the world which required industry to continuously monitor VCM in the workplace. This legislation was introduced because a link had been established between a rare form of liver cancer and prolonged exposure to VCM.

Table 3-1. Mass Spectrometer-based Air Monitoring

Sample	Source
Vinyl Chloride Monomer	PVC/Resin Production
Acyrylonitrile-butadiene-styrene	ABS Polymer Resins
Chlorinated Solvents $CHCI_3$, CCI_4, CH_2CI_2	Pharmaceutical/semiconductors
Halogenated Organic Compounds CH_3I, CH_3Br, CHF_3, C_2F_6	Petrochemical/Semiconductors
Aromatic Solvents - Benzene, Toluene, Xylene	Petrochemical/Pharmaceutical
Solvent Monitoring - Acetone, MIBK, MEK, Tetrahydrofuran	Pharmaceutical
Styrene	Fiberglass/Polystyrene Production
Epichlorohydrin	Epoxy Resins Production

Many VCM plants now have an additional requirement to monitor ethylene dichloride (EDC). The only change required is an additional calibration gas cylinder.

Acrylonitrile is used in combination with other monomers such as butadiene, styrene, and vinyl chloride to produce a range of polymers. The continuous monitoring of acrylonitrile is particularly important, since it is acutely toxic. Butadiene and styrene need to be monitored since like many other VOCs, are suspected carcinogens, and have occupational exposure limits.

A process mass spectrometer can analyze all three components in less than 15 seconds with negligible cross interference. The sample is sent through the restricted diameter of a capillary inlet or over a membrane.

The speed of analysis offered by mass spectrometers allows up to 4000 analysis to be performed and recorded each day. Mass spectrometer data systems offer features such as relay outputs to initiate visual or audible alarm systems and daily, weekly, and monthly statistical reports.

Even small leaks from valves on pipes and storage tanks can be detected. This type of analysis can indicate the degradation of valves and flanges, allows preventative maintenance before a critical leak occurs.

The membrane materials can be selected for preferential transmission of VOCs. This allows detection of compounds such as benzene, vinyl chloride monomer, and acrylonitrile down to ppb levels.

NETWORK COMMUNICATIONS

High levels of integrated control are possible by tying together distributed controllers in a communications network. This allows a reporting path that allows information to flow from one controller to another. Using coordinated control sequences, the entire building automation system can be monitored and its various functions optimized. All of this can take place transparently, behind the scenes, automatically.

Information management is the highest level of control in the networked system. Data from hundreds or thousands of I/O points in a building or building complex can be accessed quickly and used to assist in decision-making.

The appropriate communications architecture allows easy access to system information to take place at different locations throughout the facility. This access may take place at local or remote personal computer workstations.

Information management can provide both environmental compliance and energy management. Financial decision-making is also allowed along with environmental quality assurance. Automation allows a faster response to problems, as well as their resolution. Maintenance management features can also issue and track work orders.

Networked control provides quality assurance which can be used to identify, analyze and improve building operations related to both comfort and security. A large part of the building's set of plans can be loaded into the computer. For new buildings this involves access to the CAD (Computer Aided Design) system that designed the building. For older buildings the drawings can be scanned into the CAD system and then utilized by the Energy Management System.

Documentation is often needed for regulatory compliance. This documentation may include testing, proof of performance, and incidence reporting. It is essential in managing and reducing risk. Historical data can be used to identify cost-saving opportunities.

EMS EVOLUTION

The earliest Energy Management Systems used devices that were hardwired back to the computer. A distributed format evolved that used multiplexed signals over a common wire or the electrical distribution system (power line carrier systems). These power line carrier systems suffered from early reliability problems.

Multiplexing reduced the cost of wiring from remote panels to the computer but did not reduce the cost of wiring the input/output devices. Multiplexed systems have a reduced response time as the system gets larger. This response time becomes less important as distributed systems put more computing power out in the remote panels and even in the input/output devices themselves.

Direct Digital Control (DDC) evolved from the growth stage of the late 1970s which were triggered as a result of the energy price hikes of 1973 and 1977. Control system technology had been evolving but a number of factors combined to make computer-based control technology more viable. One of these was the decreasing cost of electronics which made control systems more affordable. At about the same time the interest in energy savings jumped and a number of incentives and tax credits became available which stimulated the market. These factors resulted in a demand for technology that would allow building owners to save energy.

These newly developed systems came to be known as Energy Management and Control Systems (EMCS). The computer in use at this time was the minicomputer. These systems utilized energy saving features for optimizing equipment operation, offsetting electrical demand and initiated the shut-down of equipment when not in use.

Next in the control evolution was the utilization of Direct Digital Control. This technology was used in industrial process control and even for some building applications as early as the 1950s, but it was not until much later that it became an acceptable technique for heating and cooling systems.

DDC is a closed loop control process that is implemented by a digital computer. Closed loop control means that a condition is controlled by sensing the status of the condition, taking control action to ensure that the condition remains in the desired range and then monitoring that condition to evaluate if the control action was successful.

Proportional zone control is a type of temperature control. First, the zone temperature is sensed and compared to a setpoint. When the temperature is not at the setpoint, a control action is taken to add heat or cooling to the zone. Then, the temperature is sensed again.

A thermostat with electronic or digital circuitry provides the necessary interface for digital control. It may be a microprocessor-based device that implements a sophisticated control loop, or sequence, and is capable of communications over a local area control network (LACN). The control can go beyond basic proportional temperature control and include integral or derivative control.

In this case, the integral or derivative is used to calculate the amount that the temperature is from the setpoint. The control action is limited to avoid overshooting the setpoint and the oscillations that cause delays in control response. These delays often occur with proportional control. Derivative control is often used with dynamic applications such as pressure control. Derivative control will measure the change of speed in the controlled condition and adjust the action of the control algorithm to respond to this change. The use of a Proportional, Integral and Derivative (PID) control loop allows the control variable to be accurately maintained at the desired levels with very little deviation.

Other combined sequences like PID can be used to integrate the control of several pieces of heating and cooling equipment to provide more efficient and seamless operation. Combining this type of more accurate control with networking has been an important progression for building control.

In the mid-1980s when there was no shortage of oil, the absence of a national energy policy resulted in a drop in the demand for energy management systems. The slow but continuous growth of these systems led to an awareness of the benefits of computerized control. Real energy cost reductions were noticed as well as other benefits of better control. These benefits include longer equipment life, more effective comfort levels and expanded building information. The communication features available through these systems allowed improved building management and quicker response times for problem resolution. Intelligent response is a function of system communication and provides the ability to remotely diagnose a problem. These features go well beyond energy management and are very desirable in rented properties.

Heating and cooling controls will be driven by higher energy costs and potential energy crises. There will also be a return to growth in the use of Demand Side Management.

Newer lower cost systems will drive the cost-effective replacement of conventional controls. There will also be a growth trend in performance contracting. The growing requirements of indoor air quality and related environmental requirements will open up more applications for intelligent buildings and the control integration that they utilize. Technology advancements in microprocessors, software, electronics and communications will drive growth in distributed control.

DISTRIBUTED CONTROL

Distributed control systems function as a group of control related devices and a common communication network. Building Automation System (BAS) and Building Control System (BCS) are used interchangeably. Both of these terms refer to equipment that provides building wide control, and integrates that control through distributed devices. Distributed control devices rely on networking.

A distributed control system might control heating and cooling equipment and other loads such as lighting. Distributed control is applied at each piece of equipment to provide application specific control.

Closed loop control is accomplished by monitoring the control status conditions and then executing the required control actions. One architecture uses four levels: sensor/actuator, distributed controller, building wide host and central operator interface. Other architectures combine the host and operator interface functions.

A number of products have been introduced that use a type of communication network known as sensor or field buses. This technology has been growing quickly. Remote support can take place through a modem interface over telephone lines through the Internet.

Using building wide controllers that support plug-and-play and objects, the system stores all critical system information at the controller level. Intelligent controllers of this type make it possible to dial into a system from a remote location, upload from the controllers and have full access to the system.

One trend at every layer of the control architecture has been peer to peer control. This technique distributes the critical network functions to multiple controllers and improves the integrity and reliability of the system. This type of controller can be complex, but is made possible due to the advances in electronic and microprocessor-based technology that provide greater functionality at less cost. Microprocessor technology is at the very heart of building wide and individual control advances.

Another related building wide control trend is integration at the functional level. This trend also includes a movement toward integrated control between systems with different functions such as security and building control systems.

Building wide control also means easier to use programming and monitoring systems and PC-based central operator interfaces with simple user interfaces. Peer controllers can be used for continuously interrogating the network for sequences such as morning warm-up. This feature would have been centralized in older systems. A single condition such as outside air temperature might have been monitored, and the building wide device would make a decision on start time based on this data and a stored sequence. When start up was required, that controller would signal the start of the sequence. With integrated control of this type, each controller can make independent decisions based on building wide data as well as local controller data. This results in a more reliable and effective building control system.

BUILDING CONTROL SOFTWARE

Building control software is based upon the operation of such software as Microsoft Windows. This software has a Windows look and feel for programming and monitoring. This software is mouse driven with a point and click graphical interface. Beyond this, building systems may do alarm dial outs to pagers and telephones with voice synthesis. Expanded PC-based operator interfaces can include features like preventive maintenance and on-line diagnostics.

Control sequences include such expanded applications as start/stop of non-HVAC loads and the on/off control of lighting and other electrical equipment. In these applications there are greater requirements for control integration due to the distributed nature of the control system. At this level, as well as the building wide level, peer-to-peer controllers are common.

General-purpose controllers can provide full local control requirements and integrate with both the building wide controller and the appropriate zone level controllers to provide building wide functions. Equipment level applications are energy intensive and include air handlers, chillers and boilers. The characteristics of the control include data point monitoring and multiple control sequences such as reset and warm up.

CONTROL TRENDS

The speed of information transfer can be increased by switching from twisted pair cables to coaxial or fiber optics, however, these types of cables add to the installation costs. In the future, communications between sensors and multiplex boxes and the rest of the system may use a combination of technologies including traditional means such as twisted wire and coaxial and non-traditional methods such as infrared or radio wave.

The first control panels used individual pilot lights, then came single line light emitting diode displays. The next evolution in control interfaces came with text only, monochrome CRTs. Today, high resolution color graphics provides users with realistic images that are updated every second.

Virtual reality may allow the operator to experience the environment. Special headsets and gloves may be used. After a complaint of a hot or cold temperature or a draft, an operator may zoom in to the space to feel and measure the temperature.

Zooming inside the VAV box, the operator could check the damper position and view readouts of air volume and temperature. The thermostat or damper control could be adjusted while observing the system's operation. The operator could also check the operation of fans, boilers and chillers using this zoom control.

Adding a sensor to a room could be a simple operation. The sensor may have a self-adhesive backing and stick to the wall. Power could be supplied to the unit by a built-in solar cell with battery backup. The sensor would broadcast using infrared, radio wave, or microwave. The computer will recognize the sensor and assign a point number. The system would map the location of the sensor using triangulation of the signal and its internal map of the building. A self-optimization routine would be used to search for the optimum control strategy to utilize the new sensor.

Demand limit control is a technique that raises the cooling setpoint in order to reduce some stages of cooling. This is a building wide sequence that requires equipment turn-off and avoids demand peaks.

Power measurement is becoming easier with intelligent devices and systems can monitor, measure, protect, coordinate, and control how power is used. Power and control monitoring systems use meters, protective relays, circuit break trip units, and motor starters. They can communicate information over an Ethernet network to a central location for remote monitoring, alarming, trending, and control. Power-monitoring software can be used to analyze energy use and power quality. It can identify load profiles to help with rate negotiation. If companies know their energy profiles, how and when they consume power, they can negotiate better rates for the type and amount of power they need.

LOAD SHAPING

Load-shaping involves the prediction of demand excursions for shedding loads or starting generators to avoid setting new peaks. It is also useful in predicting maintenance requirements or providing energy bills for department or area tracking of actual power use.

Intelligent metering and monitoring systems offer a low-cost method for quickly implementing energy saving practices. A Cutler-Hammer plant in Asheville, NC, installed a power management system in early 1997 when energy bills were running close to $45,000 a month. After 6 months of installation, the plant energy saving was $40,000. The power management system allowed plant engineers to identify wasteful procedures, shift loads to level the demand and perform preventive maintenance. Better control of area lights during off hours was possible. Large electric ovens were used only during the late shifts when the total energy demand was lighter. Maintenance technicians were able to locate abnormal conditions with monitoring screens and then service the equipment before it broke down. Total return on investment was predicted to be less than two years.

INTEGRATING CONTROL AND POWER MANAGEMENT

Linking power management systems to control systems allows the power information to flow from both systems. Load profiles can be developed to find any energy inefficiencies. Energy scheduling can be used to find the optimum energy schedule for new product lines or processes.

Real-time utility pricing means that production schedule energy requirements need to be compared with energy rate schedules for optimum energy benefits. The new energy supply market requires more companies to give back energy capacity during peak energy use times by scheduling lower-energy production. This can result in significant savings.

POWER QUALITY

Power quality may have significant effects on product quality. Voltage sags or swells may result in product defects. A power management system can be linked with a quality management system to help identify power-quality-caused problems. This can reduce reject rates and improve productivity.

Power quality problems are defined as any unwanted noise, transient or sag that can disable electronic components if left unprotected. Power quality problems can come from transients from power or telephone companies or excessive signal noise or sags from machinery inside a building.

One Internet company that required maximum availability developed systems consisting of generator backup and 500-kVA UPS units for AC power. Parallel UPS units were used for redundancy and capacity. For their DC power, they used a 2400 ADC powerplant.

Monitoring power consumption can also reduce maintenance costs. Monitoring the energy a motor is drawing can show you when a tool begins to get dull and produce bad parts or when a motor is overheating. Replacements can be made as preventive actions without costly downtimes or producing defective products.

Energy controls are only part of an environmentally conscious policy. Modern civilization is electricity-based, but over the past 20 years most of the energy savings has mainly been in heating.

LIFE CYCLE ANALYSIS

Design-for-environment (DFE) tools are emerging to help in energy saving. Those techniques include life-cycle analysis (LCA), pollution-prevention guidelines and parts/material recovery and recycling analysis. LCA is an overall process which evaluates the environmental impacts from acquiring raw materials, through manufacturing and use, to the final disposing of a product. The conceptual framework for LCA moved out in the 1990s beyond that of the 1970s and 1980s.

The focus was once on constraining the discharge of pollutants into the air, water, and land. In the 1990s, the Environmental Protection Agency switched from pollution control to pollution prevention. The objective is to cut pollution using natural ecosystems as the model. Industrial systems could not be open-ended, dumping endless by-products, but closed, as nature is, cycling and recyling.

The practice of LCA stems largely from a text published by the Society of Environment Toxicology and Chemistry (SETAC). The "Technical Framework for Life-Cycle Assessment" considers the source of raw materials, dependency on nonrenewable resources, energy use, water use, transportation costs, release of carbon dioxide and recovery for recycling or reuse. See Table 3-2.

Table 3-2. Environmental Requirements/Issues

Materials - Amount (intensiveness), Type
    Direct product related/process related
    Indirect Fixed capital (building and equipment)
Source - Renewable Forestry/Fishery/Agriculture
    Nonrenewable Metals/Nonmetals
    Virgin recovered(recycled)/reusable/recyclable useful life
    Location locally available/regionally available
    Scarcity/Quality/composition/concentration
    Management/restoration practices/sustainability
    Impacts associated with extraction, processing, and use
    Residuals energy/ecological factors/health and safety
Energy - Amount (energy efficiency), Type
    Purchased process by-product embodied in materials
    Source - Renewable wind/solar/hydro/geothermal/biomass
    Nonrenewable fossil fuel/nuclear, Character
    Resource base factors locations/scarcity/quality
    Management/restoration practices
    Impacts associated with extraction, processing and use
    Materials residuals-ecological factors/health and safety,
        net energy

LCA requires three stages: taking inventory, assessing impact, and assessing improvements. Taking inventory involves using a database to quantify energy and raw-material requirements (inputs) and environmental outputs, such as air emissions, water effluents and solid and hazardous waste for the life cycle of the product. Energy inputs should take into account transformation costs (raw materials into products), transportation costs (running assembly-line conveyers), and any reduction cost (when using recycled materials). Michigan State University has developed mathematical system models quantifying the inputs.

Impact assessment requires knowing which materials, processes, or components may be toxic and their impact on the environment and health which varies according to the amounts involved. Disposable or rechargeable batteries require weighing performance (battery-charge life) against toxicity.

Some companies incorporate life-cycle costs and life-cycle cost-management calculations into LCA. Decision Focus of Mountain View, CA, uses a Generalized Equilibrium Modeling System with proprietary CAD/CAE software. Another company involved in economic modeling for DFE is Synergy International in Atlanta. Synergy provides activity models. The U.S. Air Force has developed IDEF, a computer-aided software-engineering tool, for defining complex sets of interacting activities in the life cycle of an aircraft.

LCA emerged to analyze the manufacturing of toxic chemicals, but now it even affects electronic sectors. The energy-efficient Green PC is an example. Computers account for about 5% of all commercial energy in use today, and this may soon double. LCA encloses the entire life cycle of a product from raw-material extraction to end-of-life management alternatives including landfilling, incineration, and recycling. Customer use of a product is a major contributor to smog, nitrogen oxides, acid rain, and carbon-dioxide release all stemming from a product's energy consumption. Higher efficiency converters can reduce the impact.

Switching power supplies generally operate at 75% efficiency and 95% efficiency should not be difficult to achieve. Switching power supplies are superior to older linear units, which are approximately 50% efficient and require large, expensive transformers and capacitors. Standby or idle power for some products like telephone-answering machines are greater than the power consumed during operation. A line-powered electronic clock costing $10 consumes nearly that much cost in electricity in 10 years.

The Life Cycle Center of the Technical University of Denmark has made an analysis of energy consumption for a frequency converter and a portable telephone. In the study of the portable phone, energy spent in production turns out to be greater than lifetime use. The energy expended in production included that required for not only material transformation but also the energy needed to keep workers comfortable such as heating and air conditioning.

Hewlett-Packard and Xerox are recycling their hardware in Europe. In 1992 Xerox reprocessed 50,000 field-returned copiers to yield 755,000 components (51% by weight), and recycled 46% by weight into reusable materials. This left only 3% of the parts for disposal. All plastic parts should carry recycling symbols. Making parts from fewer material types and reducing paint, platings, and screws also aid in recycling.

GREEN LIGHTS PROGRAM

The U.S. Environmental Protection Agency (EPA) launched the Green Lights program in 1991. Green Lights is a voluntary, nonregulatory program aimed at reducing the air pollution that results from electric power generation. Participants are committed to upgrade a total of 4 billion square feet of facility space, this is more than three times the total office space of New York, Los Angeles, and Chicago combined.

Green Lights Partners are public and private organizations that agree to upgrade their lighting systems wherever profitable. The test of profitability for upgrades is a return on investment of the prime rate plus 6%. Most Green Lights Partners cut their lighting bills in half, while improving their work environment.

Green Light Partners agree to survey the lighting system in all of their facilities and upgrade the lighting system in 90% of qualifying building space. The upgrades must be completed in 5 years.

More than 600 firms have signed onto the EPA program, including 35% of the Fortune 500 as well as federal, state and local governments. Schools and universities make up about 15% of the list.

UTILITY REBATES

Electric utilities are also encouraging customers to reduce their energy consumption by offering rebates to those who upgrade their lighting or other building systems. Utility rebates can significantly defray the initial cost of a lighting upgrade while shortening the payback. A lighting management company can assist in applying for the maximum rebate for which it qualifies under the utility's program.

A lighting system upgrade typically pays for itself in 2 to 3 years. After payback, the continued energy savings provide a positive cash flow.

EXAMPLES OF LIGHTING UPGRADES

Mercer University in Atlanta, GA, upgraded their Wildeby Library, a 64,000-square-foot facility. Before the upgrade the lighting equipment included: 1556 T-12 lamps, 778 magnetic ballasts, 350 incandescent lamps. The new equipment included 1556 T-8 lamps, 389 electronic ballasts and 350 compact fluorescent lamps. The annual energy savings was 66 kilowatts, a 61% reduction with an annual dollar savings of $10,670 and a utility rebate of $9,594. The simple payback after rebate was approximately 2 years.

At the University of Nebraska in Lincoln, NE, their Student Union is a 24,408 square feet facility. The equipment used before upgrade included: 214 incandescent lamps, 85 incandescent lamps, 350 incandescent lamps. The upgraded lighting equipment included 160 compact fluorescent (27-watt), 85 compact fluorescents (9-watt), 51 compact fluorescents (30-watt). The annual energy savings was 42.5 kilowatts, an 86% reduction. The annual energy savings was $8,458 and simple payback was approximately 9 months.

POWER MANAGEMENT

Power management may involve devices that regulate the on and off times of selected loads, such as fans, heaters, and motors. These devices reduce the electrical demand (kilowatts) and regulate energy consumption (kilowatt hours). Power management devices can be electromechanical, electronic, or computer based. The operation of one or more loads is interrupted by the power management system based on control algorithms and building-operating parameters, such as temperatures, air flow, or occupancy. The savings in electrical energy use and cost range from 0 to 50% or more.

Electrical demand is defined as the average load connected by a user to an electrical generating system. It is measured over a short, fixed period of time, usually 15 to 30 minutes. The electrical demand is measured in kilowatts and recorded by the generating company meter for each measurement period during the billing month. The highest recorded electrical demand during the month is used to determine the cost of each kilowatt hour (kWh) of power consumed.

LOAD SHEDDING

Some power management devices are known as load shedders. They reduce the demand or average load in critical demand periods by interrupting the electrical service to motors, heaters, and other loads for short periods. Since the load which has been turned off would normally have been operating continuously, the overall effect is to reduce the average load or demand for that period of time. The instantaneous load when the load is operating remains the same.

If the period involved has the highest monthly demand, significant savings are possible in rate reductions. In periods other than the highest demand period, some energy is still saved. Before the era of high energy costs, load shedding was used mainly to avoid demand cost penalties. Now, it is used to limit energy consumption, by cycling loads on and off for brief periods, as well as to reduce demand. Other techniques used to limit energy use include the computer optimization of start times, setpoints, and other operating parameters based on the weather, temperatures, or occupancy.

Power management devices need to base their control actions on the demand level in each demand period. Some units get two signals from the electrical utility meter. These signals are usually momentary contact closures of about 500-ms duration which indicate the beginning of each demand period or the consumption of one kWh of electrical energy. The meters are modified by the utility company for about $500 to $1000. External clock mechanisms or timing signals can also be used.

The energy (kWh) consumption pulse value is converted to a unit of demand (kW) by dividing it by the demand period length in hours.

$$\frac{\text{kWh per Pulse}}{\text{Demand Period in Hours}} = \text{kW (Demand)}$$

The power control device acts as a timer or time clock. Older electromechanical timers or clocks used a small electrical motor and an arrangement of cams and switches. The switches allowed several loads to be operated in a cyclical mode by turning them on or off at preset times.

Electronic timers or clocks use semiconductor timer chips and switches to perform the same functions. Popular chips like the 555 and 556 can perform these functions and cost less than 50-cents. Applications include pool heaters, signs and outdoor display lighting.

Electric energy use is limited by having the timer turn a load off and then on for brief time intervals during each demand period. This has the effect of limiting energy consumption and reducing demand.

Electronic demand limiting includes devices that monitor and measure the actual demand and provide control actions to limit the operation of attached devices when the measured demand reaches a specified value. These devices require two signals, the kilowatt hour (kWh) or demand pulse, which indicates the units of electrical energy consumed and a timing pulse, which indicates the end of one demand pulse and the start of the next one.

The demand pulse starts a counter and as these counts accumulate within a demand period, they are compared to a target value which corresponds to the maximum desired demand. When the value accumulated in the counter exceeds the target value, the attached load is switched off.

If the demand still exceeds the demand target, a second load is turned off. This is done until all available loads are turned off or until the demand no longer exceeds the target.

Once it is turned off, the load stays off until the end of the demand period. A reset pulse from the meter restores the counter to zero and turns the load back on again.

Some load shedders use a demand target that is not fixed but increases at a steady rate. Other devices allow the off-on setpoints to be adjusted independently for individual loads.

COMPUTER-BASED DEMAND LIMITING

Computer technology allows a wider range of control options. The computer memory and decision-making capability are utilized to provide customized demand programs. The computer receives inputs from the electric meter or power sensors monitoring the critical loads.

Loads may be cycled based on the maximum demand target, time of day and day or week, rate of demand increase, heating and cooling temperatures, pressures, fuel flow and rates, occupancy schedules, inside and outside temperatures, humidity, wind direction and velocity and combinations of the above factors. Durations can be variable and changed automatically according to these parameters.

A secondary demand target may also be used. If the primary demand target is exceeded, other control actions are applied. The demand target can float up or down according to demand rates, time of day, and other parameters.

Different targets and control strategies can be utilized for different sections of buildings and plants. Reports and graphs showing the consumption by demand period and by day, week, or month are available.

In air conditioning systems, intake and exhaust dampers can be controlled on the basis of air temperatures, so that the mix of air requiring the least energy is obtained at all times. The start-up and shut-down of air conditioning, heating, and lighting systems can be regulated according to inside and outside temperatures as well as occupancy to produce the conditions which consume the least energy.

DEMAND SIDE LIMITING

Utility programs for energy conservation have involved demand-side management (DSM). These programs try to impact how customers will use electricity. One strategy is to even out the demand for electricity so that existing power generating stations are operating at efficient capacities throughout any 24 hour day rather than peaking up during business hours and late afternoon and then dropping down later in the evening. The other part of DSM is to restrain the need for new electricity capacities.

These DSM energy and load-shaping activities are implemented in response to utility-administered programs. There may be energy and load-shape changes arising from normal actions of the marketplace or from government-mandated energy-efficiency standards.

In the late 1980s, utilities began offering commercial rebate programs for DSM. Some utilities pay 30 to 50% of the installed cost, while others base their rebate programs on the peak-kilowatt-demand savings achieved by new equipment. DSM programs consist of planning and monitoring activities which are designed to encourage consumers to modify their level and pattern of electricity usage.

In the past, the primary object of most DSM programs was to promote cost-effective energy usage to help defer the need for new sources of power, including generating facilities, power purchases, and transmission and distribution capacity additions. Due to the changes that are occurring within the industry, electric utilities are also using DSM as a way to enhance customer service.

DSM involves peak clipping, strategic conservation, valley filling, load shifting, strategic load growth and flexible load shaping. It may include interruptible services or curtailment of services for specified time periods for commercial customers. Peak clipping refers to reducing the customer demand during peak electricity use periods. This is done by using some form of energy management system.

Energy conservation is often rewarded by utility rebate programs. It can include energy audits, weatherization, high-efficiency motors, Energy Management, DDC systems and HVAC systems and equipment.

Valley filling increases the electricity demand during off-peak periods, which allows the utility to use its power generating equipment more effectively. Load shifting is like valley filling, since it uses power during off-peak periods. Both valley filling and load shifting programs can involve power or thermal storage systems.

Load growth planning is a related DSM program that encourages demand during certain seasons or times of the day. Flexible load shaping modifies the load according to operating needs and can result in interruptible or curtailment rates for customers.

Consolidated Edison's program involves organizations that can reduce their summer electricity bills without buying new equipment. During the summer months, these customers agree to reduce electric demand by at least 200 kilowatts on demand. More than 100 organizations were involved in this program. Duquesne Light Company in Pittsburgh and Georgia Power also have interruptible economic development rates that operate similarly.

Con Edison also offers programs with energy audits and rebates for efficient lighting, steam air conditioning, gas air conditioning, high-efficiency electric air conditioning, cool storage and high-efficiency motors.

Houston Lighting & Power (HL&P) has a program to encourage the use of cool storage technology. It provides building owners with a $300 cash incentive for each kilowatt reduction in peak demand. There is also a cool storage billing rate, which defines the on-peak demand as noon to 7 p.m. Monday to Friday throughout the year. Many buildings have increased in value and marketability as a result of these cool storage programs. In the Dallas/Fort Worth area, Texas Utilities has more than 135 cool storage systems in operation.

Kraft General Foods and Boston Edison have an Energy-Efficiency Partnership that has reduced the ice cream manufacturer's cost dramatically. This project decreased the cost of producing ice cream by one third. The ice cream manufacturer was able to upgrade most of its electrical energy-consuming capital equipment and obtain substantial rebates for the energy saved. The rebates returned more than 85% of a $3 million investment. This included refrigeration and defrosting equipment, lighting installation and monitoring equipment.

Georgia Power has its Good Cents building program for commercial customers with HVAC and lighting rebates, along with energy audits. Georgia Power has also developed an indoor lighting efficiency program.

Besides rebates there are low- or no-interest equipment loans, financing, leasing and installation assistance and assured payback programs. Wisconsin Electric Power Company offers rebates of up to 50% of the project cost and loans with multiple rates and terms for 3 to 7 years. These programs are available to building owners and managers who install energy-efficient lighting, HVAC systems, window glazing, high-efficiency motors or building automation systems.

DIRECT DIGITAL CONTROL PROGRAMS

About a third of all utilities offer rebates for controls. These controls include timeclocks, lighting controls and Energy Management Systems. An Electric Power Research Institute (EPRI) survey found that the savings from these systems were 15% of the building energy use. Paybacks ranged between 1.5 and 3 years for control system projects including DDC systems.

Commonwealth Edison Company in Chicago offers its Least Cost Planning load reduction program. In this program, businesses cooperate to curtail or reduce their electricity consumption to prescribed limits when the utility requests it. They are compensated with a special electricity rate that is performance-based. The worst performance during any curtailment period becomes the base for electricity charges.

DSM TRENDS

Increasing competition among electric utilities may curtail DSM programs in the United States. A number of utilities are cutting back on DSM program budgets while others are shifting the focus of their DSM programs to minimize the impact on electric rates.

According to the Edison Electric Institute (EEI), DSM programs grew from 134 in 1977 to nearly 1,300 by 1992. These DSM programs deferred more than 21,000 megawatts (MWs) by 1992.

In 1997, about 1,000 electric utilities had DSM programs. A little more than half of these are classified as large and the rest are classified as small utilities. Large utilities are those that produce more than 120,000 megawatt hours. This group of larger utilities account for about 90% of the total retail sales of electricity in the United States.

Rebate and incentive programs are being modified to reduce utility costs, increase participant contributions and increase program cost effectiveness. Rebates are being used more sparingly and tend to focus on one-time energy efficiency opportunities. These programs are skewed to customer classes that are underserved by other energy efficiency programs and energy efficiency measures that are difficult to promote through financing, education and standards programs.

BUILDING AND
EQUIPMENT STANDARDS

Utilities are supporting adoption and implementation of stricter building codes and equipment efficiency standards. Appliance and equipment efficiency standards are having a notable impact on electricity demand in the United States. Standards have lowered national electricity use by 3%. Some energy efficiency measures, such as power-managed personal computers, have been widely adopted without financial incentives or much utility involvement.

INTELLIGENT BUILDINGS

The increasing acceptance of energy management systems for building management applications has been pushed by federal mandates. Energy saving systems integrate the operation and management of heating, ventilation, air conditioning, security, light and fire safety systems to reduce energy costs and minimize carbon dioxide emission of commercial buildings. The weak link in most older systems is the dependence on a human operator.

The future vision is a building that almost runs itself, from adjusting HVAC loads to dimming the lights. Many existing buildings should include such integrated energy controls systems by 2005. New constructions will be designed to utilize practical, integrated building controls.

Energy efficiency is part of an overall goal to reduce energy use and carbon dioxide emissions. The result will be practical, computerized energy management systems that unify the operation and monitoring of heating, ventilation, air conditioning, security, lighting and fire safety systems. These systems will be largely self-managing, correcting changes within the building automatically and alerting building personnel when problems occur.

Most of the building control systems being designed come with built-in computer network connections. These allow the individual controls systems to communicate either directly with each other, or with a local or remote PC.

HIGH-EFFICIENCY OIL AND GAS SYSTEMS

There are newer heating system technologies that involve modifications to conventional heat exchangers or the burn design. These changes provide steady-state efficiencies approaching 90%, with seasonal efficiencies to 85%. This is about 10% better than the steady-state efficiencies of 78 to 80% for the most efficient conventional designs.

One newer technique uses spark ignition in the combustion chamber to keep exhaust gases at 120°F instead of 400°F or more. In this process almost all the useful heat is removed and the gases are cool enough to be exhausted through a plastic pipe. In this type of system seasonal and steady-state efficiencies can reach 90%.

Air and natural gas are mixed in a small combustion chamber and ignited by a spark plug (Figure 3-1). The resulting pressure forces the hot exhaust gas through a heat exchanger, where water vapor condenses, releasing the latent heat of vaporization. In subsequent cycles, the fuel mixture is ignited by the residual heat.

One system manufactured by Hydrotherm, of Northvale, New Jersey, has efficiencies of 90 to 94%. The cost of the system is between 50 and 100% higher than a conventional one, but the improved efficiency pays back the difference within 5 years.

Flue economizers are small auxiliary air-to-water heat exchangers that are installed in the flue pipe. The unit captures and recycles the usable heat that is usually lost up the flue. This recaptured heat is used to prewarm water as it returns from the distribution system. If the flue temperature is lowered too much, moisture, corrosion and freezing may occur in the flue pipe. Depending upon the age and design of the boiler and burner, a flue economizer can provide annual fuel savings of 10 to 20% and a payback of 2 to 5 years.

Air-to-air flue economizers are also available for about 1/5 the cost but these save much less energy and are usually not tied into the central heating system. They are best for heating spaces near the flue.

Conventional flame retention burners create a yellow flame, while modified flame retention burners create a blue flame in the combustion chamber. This is done by recirculating unburned gases back through the flame zone. This produces more complete burning of the fuel and results in lower soot formation. These flame systems are available as a burner

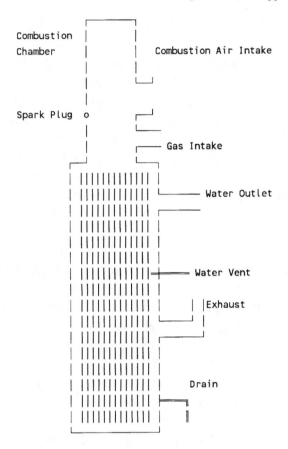

Figure 3-1. Pulse combustion boiler

for retrofit to furnaces, or as a complete burner and boiler system for hot water distribution systems.

Variable fuel flow is a technique used in burners to throttle or cut back the fuel flow rate, which reduces the flame size, as the system heating load varies. These burners have conventional steady-state efficiencies and higher seasonal efficiencies. They are available for large apartment boilers and furnaces.

There are also burners that can burn either oil or gas. They offer no efficiency advantages, but the ability to switch fuels in the event of a shortage or large price differences is their main advantage. They are available as combination burner and boiler units.

Tankless boilers offer some advantages in seasonal efficiencies, compared to conventional units, since there is less water to heat up and cool off. The savings are comparable with that of using an automatic flue damper.

ENERGY SAVING SYSTEMS

There are several technologies that are well suited to groups of buildings. These include cogeneration, district heating and seasonal energy storage systems. Cogeneration involves the simultaneous production of both space heat and electricity from an electrical generating system. A district heating system supplies heat and hot water from a central production facility to a number of residential, commercial and office buildings. A seasonal energy storage system is designed to store heat or cold energy during one season, when it is not needed, for use during another season. To be cost-effective, these types of technologies are usually applied to groups of buildings, but cogeneration and seasonal energy storage systems may be sized for small-scale applications. These technologies are not mutually exclusive since district heating may involve cogeneration or summer storage of solar energy for winter space heating.

COGENERATION

Electric powerplants produce electricity from steam that is used to rotate a power-generating turbine. The heat contained in the steam after it condenses is lost to the environment. In industrial plants, steam or process heat is often produced in the production process, but the mechanical energy in the steam or heat that could turn generating turbines is lost. Cogeneration combines the production of heat and the generation of electricity to provide a higher total efficiency than that of either process occurring separately. As the costs of fossil fuels and electricity continue to increase, cogeneration will become more attractive.

A gas turbine powerplant depends on hot, high-pressure gases produced by burning oil or natural gas. The hot exhaust gases can be used to create steam in a boiler system. The efficiency can approach 90% if the system is properly designed.

A steam turbine powerplant uses high-pressure steam produced in a boiler from burning fossil fuels or product waste to generate electricity. The low-pressure steam output can be for heating. The efficiency for this process can approach 85%.

In a diesel engine generator, waste heat can be recovered from the water-filled cooling jacket around the engine or from the exhaust gases. This heat can be used to heat water or to produce steam. Diesels often have lower efficiencies than either gas or steam turbines, but with cogeneration the total conversion efficiencies reach 90%. They are also capable of generating more electricity than comparable gas or steam turbines and are more appropriate for small-scale applications. One potential problem with diesel cogeneration is air pollution, but the newer diesel engines are cleaner than those produced in the past.

Fluidized bed combustion is a newer technology that burns coal in an efficient manner and can produce both electricity and heat. A mixture of finely crushed coal and limestone rides on a stream of air, which allows the coal to be burned at temperatures lower than conventional coal burners. This reduces the nitrogen oxide production. The limestone absorbs sulfur from the coal, which decreases the sulfur dioxide.

Cogeneration systems can also use renewable fuel sources such as wood, waste products, wood gas or methane from sewage and garbage. Sun-Diamond in Stockton, California, turned waste walnut shells into electricity for the plant and nearby homes. The walnut shells were used as fuel to produce steam to drive a turbine generator. The low-pressure steam output was then used for heat as well as to refrigerate the plant. The Sun-Diamond cogeneration system produced about 32 million kWh of electricity per year. It only used 12 million and sold the surplus power to Pacific Gas and Electric Company.

Small-scale cogeneration units are those in the 5- to 20-kilowatt range. The Fiat Motor Company has developed its TOTEM (Total Energy Module) using a four-cylinder automobile engine that burns natural gas and can be adapted to other fuels, including liquid petroleum gas (LPG) and alcohol.

It has a heat recovery efficiency of about 70% and an electrical generating efficiency of about 25%. The heating efficiency is similar to a conventional heating system, but since the unit also generates electricity its total efficiency is aver 90%. The 380 volt, 200 amp asynchronous generator unit can produce 15 kilowatts of electrical power and heat a 4- to 10-unit apartment building. Major maintenance is needed after every 3,000 hours of operation, or about every few years. This is the overhauling of the automobile engine. However, the system is cost-effective even with this overhaul requirement.

Large cogeneration units have had a long and successful operating history and are more durable than small-scale units. In smaller cogeneration units, more heat is supplied than can be used, so these systems may also include heat storage components.

Units that produce 50 to 100 kilowatts can heat multi-dwelling apartment buildings. They are fueled by natural gas or diesel fuel. Units of 200 to 2,000 kilowatts that operate on fuel oil or diesel fuel are suitable for large apartment buildings or small district heating systems.

The larger systems operate at about 35% electrical conversion efficiency and 45% heat conversion efficiency. This means that 80% of the energy in the fuel is converted to heat or electricity.

Large-scale cogeneration can provide district heating in different ways. The heat from a cogeneration unit can be used as a heat pump source, with electricity from the unit powering the heat pumps.

Attitudes and policies do not generally favor small power producers. Electric utilities are in the business of generating and selling electricity. They tend to view small power producers as competitors and have established rate structures that tend to discourage independent power generation. The Public Utilities Regulatory Policy Act (PURPA), which does not cover certain diesel engines, requires utilities to buy surplus power from and to supply back-up power to small power producers and cogenerators at nondiscriminatory fair rates.

Getting permission from the utility to install the system is becoming easier. A new net metering law in Oregon (HB3219) states that the utility has to allow grid inter-tied systems. Just after the law had been passed, the Oregon Trail Electric Cooperative (OTEC) recommended two meters for a photovoltaic system at the Grant County Fairgrounds. The planning department declared the solar installation to be a branch circuit since it did not have a subpanel.

Originally OTEC proposed administrative and inspection fees of

U.S. $131 annually for net metering customers over and above the base fees and power in/out. This would make net metering a loss to almost all customers with the 25 kW-or-under systems allowed by the law.

OTEC's concern was that distributing the cost of administration, which means hand billing in case of negative meter readings, among all their members would amount to a subsidy of the net metering customers, and raise the bills of all their customers. After a public meeting on net metering issues, the extra fees were dropped.

Cogeneration equipment must be safely connected to the utility grid. Utilities object to independent power generation by arguing that safety hazards can exist for their workers if independent systems continue to operate during system-wide blackouts. Such problems can be avoided by installation of appropriate, standard safety equipment at the cogeneration site.

A cogeneration system may use different fuels including natural gas, residual fuel oil, heating oil, diesel fuel and gasoline. Alternate fuel sources also include coal liquids or wood gas. Large-scale systems may be more cost-effective and preferable to smaller ones. If a system is properly sized and installed, it will cost less per unit of energy produced.

If multiple, smaller units are used, at least one of the units can be operating continuously, providing electricity at all times.

If some of the electricity generated is used for space heating, the system can be downsized by about 1/3. If the electricity is used to power water source heat pumps, an even smaller system is required.

REGULATORY ASPECTS

A cogeneration unit may fall under the provisions of one or more environmental and regulatory acts that cover power generation and industrial installations. Most systems of 5 to 100 kilowatts are likely to be exempt from environmental regulations except local building and zoning codes. Larger systems with a capacity in the area of some 500 to 2,500 kilowatts must comply with emission limits for five pollutants: nitrogen oxides, sulfur dioxide, small suspended particulates in the air, carbon monoxide and the photochemical oxidants found in smog. State regulations may also apply to small cogeneration systems. Regulations that affect small cogeneration systems include those governing noise pollution, water discharge and solid waste disposal.

Systems with a generating capacity of 75,000 kilowatts or less are exempt from most federal regulations governing power generation. Systems larger than about 75,000 kilowatts, or that sell 25,000 kilowatts or 1/3 of their generating capacity must comply with the Environmental Protection Agency's Stationary Sources Performance Standards for Electric Utility Steam Generating Units.

In densely populated areas, a large cogeneration system may be required to comply with emission standards and install pollution control technology. There may also be noise pollution standards and water, air discharge and solid waste disposal permits.

DISTRICT HEATING

District heating involves supplying hot water for space heating and hot water use from a central production facility to a group of residential or commercial buildings. District heating networks in Europe serve large portions of the populations of some countries. In Sweden, 25% of the population is served by district heating, in Denmark the number is over 30%, in the Soviet Union and Iceland it is over 50%. In the United States, district heating serves only about 1% of the population through older steam supply systems. In Europe, many of the district heating systems were installed during the rebuilding that followed World War II.

The main advantage of district heat is that it replaces relatively inefficient home heating systems with a more efficient, centralized boiler or cogeneration system. This offers the possibility of serious energy savings, although some heat is lost during the distribution of hot water.

District heating systems can use the waste heat from electric generation and industrial plants that would be released to the air or to nearby water supplies. Some estimates suggest that district heating could save as much as one billion barrels of oil per year in the United States.

A centralized boiler or cogeneration system can be used to produce heat. Large, centralized oil-fired boilers can remove as much as 90% of the energy contained in the fuel. Cogeneration systems can have a total heat and electricity efficiency approaching this.

In some European cities, waste heat from fossil fuel electric powerplants is used for district heating with an overall energy efficiency of 85%. These powerplants were not originally constructed as cogenerating units. Waste heat from industrial process plants can also be used. Geothermal sources are used to provide heat for district heating systems in Iceland and Boise, Idaho.

Hot water can be transported over longer distances with little heat loss while steam heat distribution systems can only serve high-density regions. The largest steam system in the United States is part of New York's Consolidated Edison Company which serves a small part of Manhattan Island. The larger pipes or mains carry 200 to 250°F water under pressure. Return mains carry the cooler, used water at 120°F back to the central facility.

Costs can be reduced with the use of new types of pipes, insulating materials and excavation techniques. Plastic piping in long rolls is laid in plastic insulation and placed in narrow trenches. Using these techniques, hundreds of feet of pipe can be laid quickly. Metal radiators can also be replaced by plastic units.

District heating systems are often financed by municipal bonds at low interest rates, to be repaid over a 30- to 40-year period. This makes the annual cost per home competitive with or less than that of conventional heating systems.

SEASONAL ENERGY STORAGE

A seasonal energy storage system is designed to store natural heat or cold during one season, when it is not needed, for use during another season. These systems have a large energy storage component. They collect essentially free heat or cold when they are plentiful and save them until required. The only energy consumed is that needed to run the various parts of the system. Three types of systems exist: annual cycle energy system, integrated community energy system and annual storage solar district heating. The first two can provide both heating and cooling. The third is used for heating only.

The annual cycle energy system (ACES) has two basic parts (Figure 3-2); a very large insulated storage tank of water and a heating-only heat pump. The tank contains coils of pipe filled with brine (salt water) warmed by the water in the tank. The brine circulates through a heat exchanger and transfers its heat to the heat pump refrigerant.

During the heating season, heat is removed from the water tank by the brine and transferred to the building at a temperature of 100 to 130°F. The system may also be used to provide domestic hot water. As heat is removed from the tank, the temperature of the water drops below the freezing point and ice begins to form on the brine circulation coils. By the end of the heating season, ice fills the entire tank.

This ice is then used during the summer to provide chilled water for air conditioning. While the ice remains in the tank, the only power required for cooling is for the operation of a circulator pump and a fan.

The annual coefficient of performance (COP), which is the ratio of

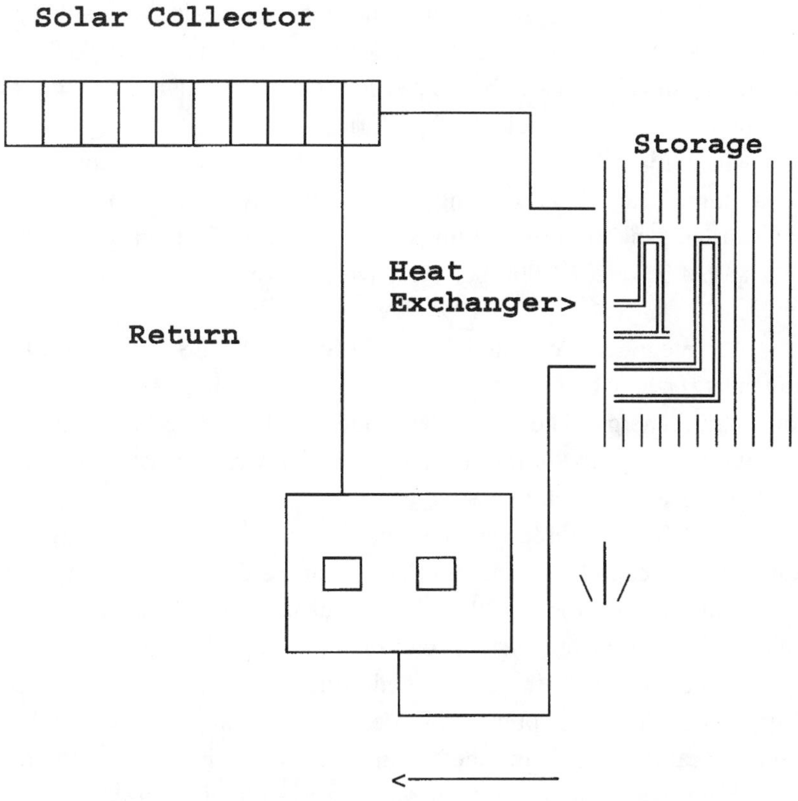

Figure 3-2. Building ACES

heat produced to energy consumed, for heating and cooling in these systems can approach 3.5. For the cooling cycle alone, the COP can be as high as 12.

The system is usually sized to meet the summer cooling requirements, rather than the winter heating load, of a building. In order to meet the total heating requirements of a building, an ACES is best suited for climates where the heat provided to the building from the tank during the winter is nearly equal to the heat removed from the building for cooling and transferred back into the tank during the summer. This is possible in areas where the winter and summer climates are not too extreme, such as Maryland and Virginia.

In several installations these system have been shown to use about 45 to 50% of the electricity consumed in a similar house with conventional electric resistance heating. It is more efficient than a conventional air-to-air heat pump system, since the heat source is maintained at a constant, known temperature. In moderate cold climates with 6,000 degree-days, an ACEs uses about 25% less electricity than a conventional heat pump with a coefficient of performance of 1.5.

The initial cost of an ACES is much higher than that for conventional home heating and cooling systems, mainly because of the cost of the storage tank. Energy savings in a house with electric resistance backup can be over $1,000 per year, which gives about a 10- to 15-year payback.

An Integrated Community Energy System (ICES) is a type of district heating and cooling system that uses heat pumps to collect and concentrate energy. The use of heat pumps allows free heat that would otherwise be lost to be removed from boiler waste heat, groundwater, lakes, solar and geothermal sources.

An ICES has three major components: heat pumps, a large heat source which can also act as heat storage and a distribution system. The heat pump section of an ICES may be centralized, distributed or cascaded. In a centralized system, one or more large heat pumps are used in a manner similar to the centralized boiler of a district heating system (Figure 3-3). The heat pumps are located in a central facility, and they remove heat directly from a heat source. This heat is then used to warm distribution water, which is then pumped to individual buildings.

In a distributed system, small heat pumps are located in each build-

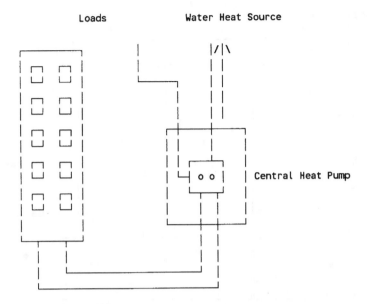

Figure 3-3. Centralized Integrated Community Energy Systems (ICES)

ing (Figure 3-4). Water from the heat source is sent directly to an individual heat pump. Heat removed from the distribution water is then used to warm the building. Some heat pumps may be used to also provide cooling.

A cascaded system uses both centralized and individual heat pumps (Figure 3-5). A central heat pump removes low temperature heat from the primary source and adds it to the distribution water, which is sent to individual buildings. Heat pumps in the buildings then use this distribution water as a secondary heat source. This system is used when the primary source water is too corrosive, such as salt water, or contaminated, such as waste water.

The distribution system of an ICES is the same as that of a conventional district heating system. Each ICES has warm water supply and cool water return mains. Systems that supply both heating and cooling at the same time may have independent distribution systems for hot and cold water. Distributed systems using groundwater as a heat source may have only a distribution water supply line. Cascaded and distributed ICESs have separate heating distribution systems for each building.

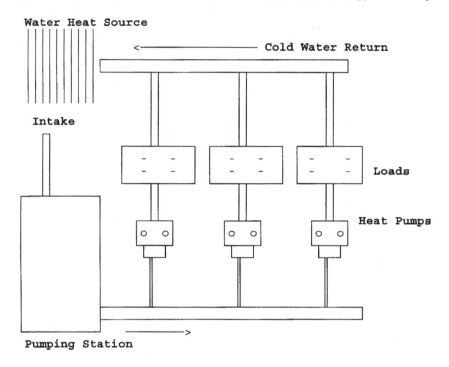

Figure 3-4. Distributed ICES

The heat source in an ICES is usually a large body of water that provides a concentrated source of heat at a fairly constant temperature. Depending on the winter climate, the heat source can be a lake, reservoir, underground storage tank, aquifer (underground river or lake), solar-heated water, sewage or waste water, geothermal energy or waste heat from industrial or commercial facilities.

In an ICES that serves both small and large buildings, the surplus internal heat from the large buildings can be used to provide source heat to smaller ones. An ICES in areas with moderate winter temperatures may use air as a heat source. Systems that use lakes or reservoirs depend on the natural collection of heat by these water sources throughout the year.

An ICES supplying simultaneous heating and cooling may have short-term hot and cold storage to hold that produced during the course of a day. An aquifer can sometimes be used to store warm water if the ground water flow is slow enough.

Cold Water Return

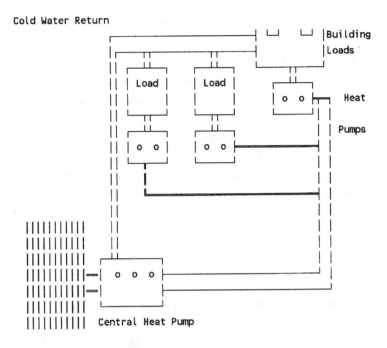

Water Heat Source

Figure 3-5. Cascaded ICES

The operation of an ICES depends upon the nature of the heat source and if the system is centralized, distributed or cascaded. If the system is centralized with surface water (lake or reservoir) used for a heat source, the heat is removed directly from the source and transferred through the distribution system. Otherwise, source water is pumped directly to buildings for use by individual heat pumps. This type of heat source is used only in warmer climates or with very large sources.

When an aquifer heat source is used, two wells are required. One is used for groundwater return. The ground water is distributed to individual heat pumps and then returned to the aquifer. If the aquifer is large enough, pairs of wells can be drilled throughout the community.

In a groundwater system that supplies both heating and cooling, the aquifer can be used to store heat or cold. If warm water is injected and withdrawn only from the upper part of the aquifer and cool water is injected and withdrawn only from the lower part, the slow movement of water in the aquifer and the difference in temperature will tend to keep the warm and cool layers separated.

Many geothermal sources of water have a temperature range of 70 to 90°F. Such sources exist in many parts of the United States. An ICES using a geothermal source would probably be cascaded, with a large centralized heat pump and additional heat pumps in each building. This would be necessary since mineral deposits from geothermal sources could block pipes and pumps.

Heat is removed from the source water and added to the distribution water by the central heat pump. The individual heat pumps then use this distribution water as a source of heat for individual buildings.

Solar energy can be used to warm heat pump source water. In this system solar collectors are mounted on a large, insulated water tank where the warmed water is stored. Most of the heat is collected in the summer for use during the winter.

In the winter, the hot water can be used directly for space heating until it cools to about 85 to 90°F. The remaining heat can be removed and concentrated by a centralized heat pump.

An ICES using a large fabricated tank of water as a heat source can operate as a community-scale ACES. The water in the tank is slightly higher than 32°F. During the winter a centralized heat pump removes heat from the tank, causing the formation of ice. This ice is then used for summertime air conditioning or for winter cooling of large buildings.

Sewage and wastewater heat sources are usually not much colder than the buildings from which they come. A cascaded ICES can remove heat from waste water and transfer it to the distribution system which then acts as a secondary heat source for heat pumps in individual buildings.

Waste heat is often lost into the environment by industrial facilities in the form of hot water. This hot water can be used directly by the heat pumps in a centralized ICES.

ICES have several advantages over conventional district heating systems or individual building heating systems. An ICES will often serve business, commercial and residential districts. Since the peak heating and cooling demands of these different sectors may not occur at the same time of the day, a single moderately sized system can meet the varying peaks of the different sectors.

When the ICES contains a short-term heat storage component, such as a water tank, the system can operate continuously and at a steady level around the clock with peak heat demand requirements drawn from storage.

Conventional heating system burn fossil fuels at high temperatures to heat water to 120°F. District heating systems operate in the same way. In these cases, when the hot water cools to 90°F or less, it is no longer warm enough to supply heating. This remaining heat is eventually lost to the environment. An ICES can recover this low-temperature heat that would otherwise be wasted. This helps to increase system efficiency.

ICES are generally found to be economically competitive with conventional heating systems such as furnaces and/or boilers in individual buildings or district heating systems using fossil fuels. Capital costs are a good deal higher than those of conventional systems, but ICESs have lower energy requirements. Free environmental energy is substituted for the burning of fossil fuels. In some ICESs, electricity consumption is greater than in conventional systems lacking heat pumps, but the total consumption of all forms of energy is lower.

ACESs and ICESs depend on heat pumps and storage systems, and require notable amounts of energy to operate them. An annual storage solar district heating system could supply most of a community's annual space heating requirements with a minimum of nonrenewable energy.

These systems cannot provide air conditioning so they are mostly suited to northern climates. This is because over the course of a year even northern locations such as Canada receive as much sunlight per square foot as Saudi Arabia. The problem is that most of the sunlight falls in the summer when it is not needed for heating. In annual solar storage the system collects heat in the summer for use during the winter.

An annual storage solar district heating system requires a heat store, a collecting area and a distribution system (Figure 3-6). The storage can be either an insulated earth pit or a below-ground concrete tank. Both have insulated concrete covers and are filled with water. Collectors are mounted on the cover of the storage tank and are rotated during the day so they always face the sun.

During the summer, the collectors heat water for storage and for domestic hot water. During the winter, the collecting system heats water that is used directly for heating purposes. When additional heat is required, the hot water stored in the storage tank or pit is used.

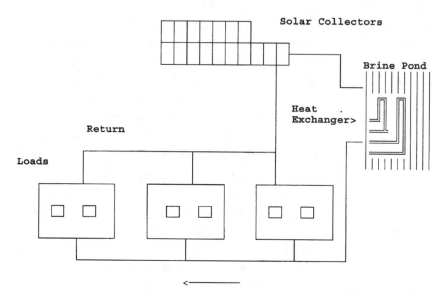

Figure 3-6. District ACES

Water is removed from the top layers of the storage tank. The cooler used water is pumped back through the collectors or into the bottom of the storage tank.

One project using a large rock cavern in Sweden provides district heating for 550 dwellings. A housing project at Herley, near Copenhagen in Denmark, uses a central solar collector and a large insulated water tank buried in the ground. Solar heat provides most of the space heating requirements of the 92 housing units. When the temperature of the heat store falls below 45°C, heat is transferred via a heat pump, powered by a gas engine, which boosts the temperature to 55°C. This process continues until the temperature of the heat store has fallen to 10°C, at the end of the heating season. Waste heat from the engine is also delivered to the heating system, and a gas boiler is used as a back-up.

In summer, the main heating system is shut down and 90% of the domestic hot water requirements of the housing units are provided by additional solar collectors on each of the eight housing blocks. All of these systems operate in latitudes far to the north of American cities. This type of system can also be implemented by a gas furnace as shown in Figure 3-7.

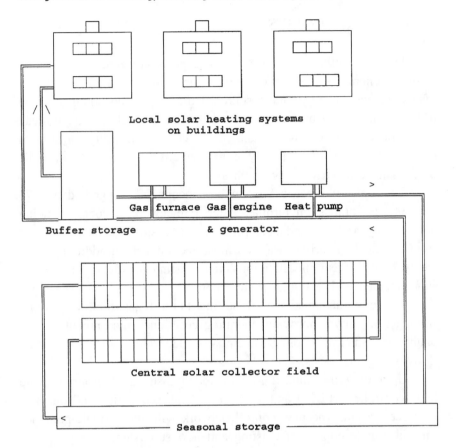

Figure 3-7. Solar district heating

An annual storage solar district heating system is capable of supplying 90% of the annual heating requirements for the homes in a community. Depending upon the climate zone, the required collector area per house can range from 70 to 300 square feet. This can be reduced if residential heat loads are lessened through increased levels of weatherization and the addition of passive solar features.

Solar district heating offers a number of advantages over conventional single-residence active systems. The collectors can be set aside in an open area and problems with access to the sun do not arise.

The heat storage capacity is not constrained by space limitations in any one building and the storage tank can be as large as necessary. Since the system is equipped for annual storage, solar collection is not dependent on the day-to-day weather conditions.

TURBINE RETROFITS

Due to declining electric utility capacity reserves, combustion turbine generators that have been virtually moth-balled during the 1980s are once again a vital resource. Utilities are depending more on turbines. Increasing unit capacity and operating reliability are critical, since not meeting a peak demand period can cost a utility millions of dollars. As a result, capacity enhancement and controls upgrade projects have taken on new emphasis.

The Lincoln Electric System of Lincoln, NE, upgraded a 1975 GE 7001-B combustion turbine. After a review of options, including both supply and demand side projects, increasing the energy output from the existing turbine was the most economical. In addition to capacity enhancement modifications, the existing analog controls were supplemented with a digital system upgrade.

The upgrade consisted of adding natural gas fuel, improving starting reliability, improving operating costs and environmental compliance. An inlet air cooling system was added along with a new high flow inlet guide vanes.

Upgrading the control system added distributed programmable logic controllers (PLCs) and a PC-based data acquisition and operator interface system. The new control network will allow single-point local and remote control of existing and new subsystems.

Both the natural gas fuel system and the inlet cooling plant required additional digital controls to augment and supplement the turbine's original Mark I analog control system. PID control of the natural gas fuel system is used with automatic transfer between gas and distillate fuel. It was desired to integrate and control the turbine inlet cooling system with existing turbine controls.

The overall control scheme would include a 23-ton/hour ammonia-based ice production plant and recording nitrogen oxide emissions as required under the state operating permit. It provides the on-line display and logging of all plant variables, including real-time, historical charting and graphics capabilities. It supplements the turbine's combustion monitor, a key turbine safety system.

Enhanced turbine operations include scheduling daily turning gear operation. The system provides an interface between plant systems and a supervisory control and data acquisition (SCADA) system used for remote operation and allows for easy expansion for future plant systems.

The system uses two Allen-Bradley PLC-5 programmable logic controllers interconnected with a Data Highway Plus local area network (LAN). Microcomputers are interfaced to the PLC network dynamic data exchange (DDE) cards.

The natural gas control system was supplied by TTI/Unitech. Microprocessor-based refrigeration compressor controls were supplied by Vilter. The Wonderware Intouch operator interface software was used.

One PLC is dedicated to combustion turbine operations while the other PLC controls the refrigeration and inlet cooling plant. The system has 160 analog points and 120 digital I/O, with three PID loops. In addition to monitoring and operating the refrigeration plant and inlet cooling system, the refrigeration PLC uses an RS-232 link to interface with the controls for a 500-ton refrigeration plant compressor.

A SCADA system allows utility system energy dispatchers to remotely start and control the turbine/inlet cooling plant. A modem-equipped, laptop PC allows operators to monitor plant operation from any location with a telephone.

The plant operators built their own screens for the real-time display of key plant variables. This allows them to identify, at a glance, anomalies in operating parameters. Comparing the on-line data with alarm setpoints, the operators are alerted to possible problems.

The system also provides additional safety features such as monitoring ambient temperature and relative humidity to alert the operators to conditions conducive to ice formation on turbine inlet structures.

The data acquisition system allows trend analysis of successive operating curves. These assist in making decisions to adjust routine inspection and maintenance periods. With the addition of the data acquisition system, Lincoln Electric is moving beyond preventive maintenance to predictive maintenance.

Diagnosis of a malfunction is greatly enhanced with accurate time line data. The data acquisition system's storage allows data to be transferred for analysis.

References

Barker, Jennifer, "Hoops and Hurdles," *Home Power*, Issue #80, December 2000/January 2001, pp. 64-65.

Boyle, Godrey, Editor, *Renewable Energy Power for a Sustainable Future*, Oxford, England: Oxford University Press, 1996.

Davlin, Tom, "PLCs Upgrade Power Plant Control System," *Intech*, Vol. 40 No 10, October 1993, pp. 38-39.

Khan, Khalid, "Mass Spectrometers, Poll Plant Pollution," *Intech*, Vol. 40 No. 10, October 1993, pp. 40-42.

Scientific Staff of the Massachusetts Audubon Society, *The Energy Saver's Handbook*, Emmaus, PA: Rodale Press, 1982.

Internet: www.web5.infotrac.galegroup.com

Lippe, Jim, "Green Strategies Cope with Electronic Products' Energy and End of Life," Nov. 10, 1994, pp. 1-15.

CHAPTER 4

POWER QUALITY, LOADS AND HARMONICS

ALTERNATING VOLTAGE AND CURRENT

Standard utility power systems supply alternating current and voltage (AC). Standard load voltages are 120, 240 or 440 volts. The frequency of these voltages is 60 cycles per second. Cycles per second is given the unit Hertz or Hz. The voltage waveform developed by an AC generator is a sine wave. The waveform repeats itself 60 times each second, and passes from a maximum positive value to a maximum negative value. The sine voltage can be defined by its peak value or its peak-to-peak value. But, in order to determine the amount of work a voltage performs, or the effect that a voltage has on a circuit or load, AC voltages and currents are measured and identified by their effective values. The effective value is the peak voltage or current divided by the square root of two. This is because

$$P = Veffective^2/R = Vpeak^2/2R$$

so,

$$Veffective = Vpeak/2$$

In a load with an electrical resistance, the AC voltage and current perform work during both the positive and the negative cycles. The power waveform acts as if both halves of the waveform are positive.

During the positive cycle both current and voltage are positive so power is positive. During the negative cycle both voltage and current are negative and their product is positive. The current and voltage waveform in a resistor are in phase. This gives

$$P = -I \times -E = +P \text{ Watts}$$

129

These are the conditions in a purely resistive circuit. The product of current and voltage equals the power in a circuit with only resistance. For any type of load with resistance, inductance or capacitance, this becomes

$$P = I \times E \text{ cosine}$$

where cosine θ is the phase angle between current and voltage. In a purely resistive load, the phase angle is zero degrees, and the cosine of zero degrees is 1.

POWER FACTOR

Solving the above relation for cosine θ gives the power factor (pf) which is the ratio of real or actual power (W) to apparent power (VA).

$$pf = \frac{\text{Real Power (P)}}{\text{Apparent power } (I \times E)} = \text{Cos } \theta$$

pf angle in degrees = $\cos^{-1}\theta$
Real Power = W = actual power = supplied to resistive loads
Reactive Power = VAR = (W)tan θ = supplied to reactive loads

Power factor may be leading or lagging real power depending on the load. Capacitors can be used to improve the power factor of a circuit with a large inductive load. Current through a capacitor leads the applied voltage by 90 degrees and has the effect of canceling the inductive lagging current. For example, the power in a 10 Ω purely resistive load in a 120 volt circuit

$$I = E/R = 120V/10\Omega = 12A$$

$$P = I \times E \text{ cosine } \theta = 12 \times 120 \times 1 = 1440 \text{ Watts}$$

INDUCTIVE LOADS

Inductance is the property of a coil that opposes a change of current in a circuit. This property of a pure inductor causes the voltage to lead the current by 90 degrees in an AC circuit. A pure or ideal inductor is one without any electrical resistance. This property has a major effect on the power. Since the current and voltage are out of phase by 90 degrees their product, the power goes from positive to negative each half cycle. This results in a positive and negative power wave which means that the inductor takes power from the source and returns the power back to the source. The result is zero power dissipation. There is current and voltage in the inductor, but no power. This condition is defined by another term called volts-amperes reactance (VAR).

The power in a pure inductor with an inductive reactance (X_L) of 10 Ω in a 120V AC circuit is

$$I = E/X_L = 120V/10\Omega = 12A$$

$$P = I \times E = 12A \times 120V = 1440 \text{ VARs}$$

An AC wattmeter connected to this load would measure zero watts since no power is dissipated, a VAR meter would measure the product of current and voltage.

CAPACITORS AND POWER FACTOR

Capacitance is a property that opposes any change of voltage in a circuit. As the capacitor voltage changes, the capacitor charges and discharges, opposing voltage changes across its plates. The current in a capacitor leads the voltage across the capacitor by 90 degrees. The product of current and voltage is a power wave that is very similar to that of the inductor. The power dissipated is zero and the current and voltage product is defined in VARs.

The resulting power dissipated in a capacitor with a capacitive reactance (X_C) of 10 Ω connected across a 120V source is

$$I = E/X_C = 120V/10\Omega = 12A$$

$$P = I \times E = 12A \times 120V = 1440VARS$$

The power dissipated by the resistive part of a load can be measured by a wattmeter. The power factor of a circuit or load can be found from the current and power of the circuit. It can also be measured with a pf meter, which is also called a VAR meter.

The current of a motor measures 10-A with a clamp-on ammeter, the voltage is 220V and a wattmeter measures 1250-W. Then,

$$\text{Apparent Power} = I \times E = 10A \times 220V = 2200VARs$$

$$pf = 1250W/2200VARs = 0.568$$

The phase angle between the current and voltage in the motor load is the angle with a cosine of 0.568. This angle is 55.9 degrees.

When a large number of inductive motors are used, the electrical system will operate with a low power factor. This means that the true power measured by the utility kilowatt-hour meter is much less than the power delivered by the utility.

The utility may charge an additional fee for the power delivered, or demand a correction of the power factor. The pf can be corrected by placing the correct value of capacitor in the line. The correct value can be calculated and most utility companies will supply guidelines for the method of selection of the capacitor and proper methods for installation.

A low power factor caused by the use of highly inductive (magnetic) devices indicates a low electrical system operating efficiency. These devices indicate:

- non-power factor corrected fluorescent and high intensity discharge lighting fixture ballasts (40%-80% pf),
- arc welders (50%-70% pf),
- solenoids (20%-50%),
- induction heating equipment (60%-90% pf),
- lifting magnets (20%-50% pf),
- small dry-pack transformers (30%-95% pf) and,
- induction motors (55%-90% pf).

Induction motors are often the major cause of low power factor in most industrial facilities since they are the main load. The correction of low power factor is a vital part of power generation, distribution and utilization.

Power factor improvements can increase plant capacity and reduce any power factor penalty charges from the utility. They also can improve the voltage characteristics of the supply and reduce power losses in feeder lines, transformers and distribution equipment.

Capacitor correction is relatively inexpensive. The installation of capacitors can be done at any point in the electrical system, and will improve the power factor between the point of application and the supply. However, the power factor between the load and the capacitor will remain unchanged.

Capacitors are usually added at each motor or groups of motors or at main services. Placing the capacitor on each piece of equipment increases the load capabilities of distribution system. There is also better voltage regulation since the capacitor follows the load. However, small capacitors cost more per kVAC than larger units. The economic break point for individual correction is generally at 10 HP.

Placing the capacitors at the main service provides the lowest material installation costs but switching is usually required to control the amount of capacitance used. If the loads contributing to power factor are relatively constant, and system load capabilities are not a factor, correcting at the main service can provide a cost advantage. If the low power factor is due to a few pieces of equipment, individual equipment correction would be most cost effective. Most capacitors used for power factor correction have built-in fusing. If they do not, fusing must be provided.

POWER QUALITY

The electronic circuits used in motor controls can be susceptible to power quality related problems. Common problems include transient overvoltages, voltages sags and harmonic distortion. These problems may surface in the form of control anomalies, nuisance tripping and in some cases circuitry damage.

Capacitors may be used in the electrical system to provide power factor correction and voltage stability during periods of heavy loading. When these capacitors are energized, transient overvoltages may occur.

Circuits may be sensitive to temporary reductions in voltage. These voltage sags are usually caused by faults on either the customer's or the utilities electrical system.

Lighting systems and other electric devices can cause distortion in the electrical current, which can affect power quality. Incandescent lighting systems do not reduce the power quality of a distribution system because they have sinusoidal current waveforms that are in phase with the voltage waveform. The current and voltage both increase and decrease at the same time. Fluorescent, HID and low-voltage systems, which use ballasts or transformers, can have distorted current waveforms. Devices with heavily distorted current waveforms use current in short bursts, instead of following the voltage waveform. This has the affect of distorting the voltage waveform.

The load current waveform will be out of phase with the voltage waveform creating a phase displacement that reduces the efficiency of the alternating current circuit. The current wave will lag behind the voltage wave.

The device produces work during the time of the circuit's active power. This is the component of power that is in phase. When the current and voltage are out of phase, this produces reactive power.

Reactive power place an extra load on the distribution system. It must have the capacity to carry reactive power even though it accomplishes no useful work.

HARMONICS

A harmonic is a higher multiple of the primary frequency (60-Hertz) superimposed on the alternating current waveform. A distorted 60-Hz current wave may contain harmonics at 120-Hz, 180-Hz and higher multiples. The harmonic with a frequency twice that of the fundamental is called the second-order harmonic. The harmonic whose frequency is three times the fundamental is the third-order harmonic.

Highly distorted waveforms contain many harmonics. The even harmonics (second-order on up) tend to cancel each other effects, but the odd harmonics tend to add in a way that increases the distortion because the peaks and troughs of their waveforms coincide. Lighting products usually indicate a measurement of the distortion percentage in the Total Harmonic Distortion (THD). The Total Harmonic Current (THC) depends on this percentage:

$$THC = \%THD \times \text{current of the load}$$

An electronic ballast might have a slightly higher % TDH than a magnetic ballast, the THC is usually less since the electronic ballast consumes less power. Most electronic ballasts for fluorescent lamps have filters to reduce the current distortion. An electronic ballast uses about .2 amps compared to almost 1/3 amp for a magnetic ballast.

CREST FACTOR

The crest factor is the ratio of peak current to the Root Mean Square (RMS) current value measured with an AC meter. Crest factor is used by lamp manufacturers to estimate fluorescent rapid-start lamp life. Most ballasts use a 1.44 crest factor. The closer a ballast operates to the 1.44 crest factor, the longer the lamp life and the better the lamp/ballast efficiency. Current from a ballast with a higher crest factor causes the higher peak currents to erode the lamp electrodes, shortening lamp life.

Power factor indicates the combined effects of current THD and reactive power on phase displacement. A device with a power factor of unity has 0% current THD and the load current is synchronized with the voltage. Resistive loads such as incandescent lamps have power factors of unity.

Normally, ballasted lamps have a power factor between 0.5 and 0.9, this is sometimes called a normal power factor (NPF). High power factor (HPF) devices have a power factor greater than 0.9. Magnetic and electronic ballasts for fluorescent lamps can be either NPF or HPF. HPF ballasts usually have filters to reduce harmonics and capacitors to reduce phase displacement.

The growing use of nonlinear loads has increased the complexity of system power factor corrections. The application of pf correction capacitors can heighten rather than correct the problem if the fifth and seventh harmonics are present.

Harmonic distortion occurs when electronic circuits introduce harmonics into the power system. These harmonics are due to the nonlinear characteristics of power electronics operations. Harmonics are components of current and voltage that are multiples of the normal 60-Hz AC sine wave. If severe enough, these harmonics can cause motor, transformer and conductor overheating, capacitor failures, misoperation of relays and controls and reduce load efficiencies. Compliance with IEEE-519 "Recommended Practices and Requirements for Harmonic Control in Electrical Power Systems" is recommended.

VARIABLE FREQUENCY DRIVES

Variable frequency drives change the speed of a motor by changing the voltage and frequency of the power supplied to the motor. In order to maintain the proper power factor and reduce excessive heating of the motor, the name plate volts/hertz ratio must be maintained. This is the main task of the variable frequency drive (VFD).

The main parts of a variable frequency drive are the converter, inverter and the controller. The converter rectifies the AC input to DC and the inverter converts the DC to an adjustable frequency, adjustable voltage AC signal. Both must be adjustable to provide a constant volts to hertz ratio. A circuit filters the DC before it is sent to the inverter. The controller regulates the output voltage and frequency based upon a feedback signal from the process. If the load is a pump, this is usually a pressure sensor.

Voltage source invertors use a silicon controlled rectifier to build a sine waveform for input to the motor. The steps used to build the waveform create harmonics that are reflected to the power source. The steps of the waveform causes current pulses that make the motor cog at low frequencies and damage to keyways and couplings may result.

Current source inverters use an SCR to control the current to the motor. This is also done with multiple steps and has the same problems with cogging and harmonics as the voltage source systems. Many vendors only offer the voltage or current source systems for large horsepower motors of over 300 HP.

Pulse width modulation (PWM) units have become more popular recently. Units are available from the smaller HP sizes all the way to 1500 HP. A PWM inverter produces pulses of varying widths which are combined to build the required wave form. A diode bridge is used in some converters to reduce harmonics. PWM produces a current waveform that more closely matches a line source, which reduces undesired heating. PWM drives have an almost constant power factor at all speeds which is close to unity. PWM units can also operate multiple motors on a single drive.

In a basic DC-link variable-frequency motor controller, the input AC power is converted to DC, filtered, and then converted to variable-frequency DC by an inverter. A set of SCR switches are used to convert the DC to three-phase AC power to drive an induction motor. Bypass diodes are needed for reactive power flow and to clamp the voltage to that of the DC supply. The filter supplies a DC voltage to the invertor that is largely independent of load current due to the filter capacitor. The inductor tends to keep the current constant. The AC-to-DC converter output may be fixed or variable (voltage or current), depending on the type of inverter and filter used.

In a square-wave inverter, each input is connected alternatively to the positive and negative power-supply outputs to give a square-wave approximation to an AC waveform at a frequency that is determined by the gating of the switches. The voltage in each output line is phase shifted by 120° to provide a three-phase source.

The switches produce a stair-step voltage for each motor phase. At frequencies below the rated frequency of the motor, the applied voltage must be reduced. Otherwise, the current to the motor will be excessive and cause magnetic saturation. This comes from Faraday's law:

$$\underline{V} = j\omega n\Phi$$

where ω is the electrical frequency.

A decreasing voltage level to keep the peak flux constant can be done with the square-wave inverter decreasing the DC voltage as motor speed is reduced below rated speed. This can be done by a controlled rectifier, but this produces problems with harmonics in the power system supplying the controller.

Harmonics in the square-wave inverter have two sources. At the input, the controlled rectifier generates harmonics that produce electrical noise in the power system. These can be filtered, but this reduces efficiency and the power factor, which is already low in a controlled rectifier.

The output waveforms also produce serious harmonics. The stair-step output waveforms have only odd harmonics. The third and ninth harmonics cause no problems, since they are in phase and cancel at the input to the wye-connected motor. The rest of the harmonics, mainly the fifth and seventh, cause currents that increase losses in the motor and produce no torque. These harmonics are filtered some by the inductance of the motor. The combination of these problems has resulted in the use of pulse-width modulation (PWM) systems. These systems use a more complicated switching scheme.

PWM cuts up the square wave to control the fundamental in the output and shifts the harmonics to higher frequencies that are more easily filtered by the inductance of the motor.

One way for controlling the switches in PWM is done by mixing a triangular wave with a sine wave. The triangle wave is the carrier, and its frequency is much higher than the electrical frequency applied to the motor. The sinusoidal waveform acts as a modulating waveform that controls the amplitude and frequency of the fundamental applied to the motor.

The harmonics are shifted to frequencies near the carrier frequency which are much higher than those with the square wave inverter. The higher harmonics are filtered by the inductance of the motor and the resulting current is sinusoidal with a small amount of ripple.

In the other phases of a three-phase inverter, the modulating sinusoids are phase shifted by 120° and 240°. This allows the inverter to provide three-phases of controlled frequency and amplitude.

Another method of PWM switching compares the output current with a reference signal. When the output current drops below the reference signal, the output is switched to the positive dc bus, and when the output current rises above the reference signal, the output is connected to the negative dc bus. This method controls current and torque directly and is not sensitive to variations of the motor impedance with drive frequency.

Variable frequency drive systems can result in energy savings. These savings are achieved by eliminating throttling and friction losses affiliated with mechanical or electromechanical adjustable speed technologies. Efficiency, quality, and reliability can also be improved. The application of VFD systems is load dependent.

The most common types of adjustable speed loads are variable torque, constant torque and constant horsepower loads. A variable torque load requires much lower torque at low speeds than at high speeds. In this type of load, the horsepower varies approximately as the cube of speed and the torque varies approximately as the square of the speed. This type of load is found in centrifugal fans, pumps, and blowers.

A constant torque load requires the same amount of torque at low speed as at high speed. The torque remains constant throughout the speed range, and the horsepower increases or decreases as a function of the speed. A constant torque load is found in conveyors, positive displacement pumps and certain types of extruders.

A constant horsepower load requires high torque at low speeds, and low torque at high speeds. Constant horsepower loads are found in most metal cutting operations and some extruders. Energy savings are available from the non-centrifugal, constant torque or constant horsepower loads based on the VFD's high efficiency compared to mechanical systems, improved power factor and reduced maintenance costs. An application load profile, percent load versus time, can be developed with demand metering equipment. This curve can be used to determine the possible savings. The VFD must be matched to the motor and load to maintain the motor temperature requirements and the minimum motor speeds. VFD pumps are used with a valve control system which is adjusted to maintain a constant pressure in the system. VFD fans are used with damper and inlet controls.

VFDs can be successfully applied where the existing equipment already utilizes one of the older speed control technologies. They have been used to replace motor-generator sets with eddy-current clutches. Equipment that utilizes variable pitch pulleys (Reeves type drives) can be upgraded with a VFD which reduces slippage losses and can improve product quality through better control and improved reliability.

Variable volume air handlers can use VFD drive controllers to reduce the amount of energy required to supply the required amount of air to the system. Chiller manufacturers have used VFD controllers to replace the butterfly inlet valves on centrifugal compressors. This significantly reduces the power requirements of the chiller for a majority of the operating cycle.

VFDs have been used to vary the speed of centrifugal combustion air intake fans with 50-hp motors for high pressure steam boilers. The system uses an actuator to vary the amount of gas and air that enters the burner. The air is controlled with inlet dampers. The annual savings was almost 90,000 kWh and the simple payback was about 4 years without accounting for any demand savings.

A malting plant in Wisconsin used seven 550-ton chillers to provide cold water for process cooling. Three of the chillers are on all of the time and the other four run according to the plant's demand for cooling. The chillers were controlled by variable inlet vanes.

A variable frequency drive, also referred to as adjustable speed drive (ASD), could be used for one chiller that operates about 6,000 hours per year. From load and duty-cycle information it was determined that this chiller operates at 64% of its full load capacity. The savings attainable by a VFD are estimated at 26% per year or about 400,000 kWh per year. The payback is less than 5 years.

LIGHTING AND POWER QUALITY

The National Energy Policy Act of 1992 (EPACT) was designed to reduce energy consumption through more competitive electricity generation and more efficient buildings, lights and motors. Since lighting is common in nearly all buildings, it was a primary focus of EPACT. The 1992 legislation banned the production of lamps that have a low efficacy or CRL.

In some cases, lighting retrofits can reduce the power quality of an electrical system. Poor power quality can waste energy. It can also harm the electrical distribution system and devices operating on that system.

The first generation electronic ballasts for fluorescent lamps caused some power quality problems. But, the electronic ballasts available today can improve power quality when replacing magnetically-ballasted systems in almost every application. Problems can still occur in sensitive environments such as care units in hospitals. In these areas, special electromagnetic shielding devices are used.

POWER SURVEYS

Power surveys are used to compile records of energy usage at the service entrance, feeders and individual loads. These records can be used to prioritize those areas yielding the greatest energy savings. Power surveys also provide information for load scheduling to reduce peak demand and show the characteristics of loads that may need component or unit replacement to reduce energy consumption. This involves the measurement of AC power parameters.

When using polyphase motors, watt or VAR, measurements should be taken with a two element device. The power factor is then determined from the readings of both measurements. When variable speed drives are involved, make the measurements on the line side of the controller.

MOTOR LOADS

Electric motors need to be properly loaded. There are several ways to determine motor loads. The most direct is by electrical measurement using a power meter. Slip measurement or amperage readings can be used to estimate the actual load.

Most electric motors are designed to operate at 50 to 100% of their rated load. The optimum efficiency is generally 75% of the rated load since motors are usually sized for their starting requirements.

In air moving equipment, the performance of fans and blowers is governed by rules known as The Affinity Law or The Fan Law. The first rule states that the fan capacity in cubic feet per minute varies as the fan speed (RPM). This is expressed as the ratio between fan 1 and fan 2:

$$\frac{CFM(2)}{CFM(1)} = \frac{RPM(2)}{RPM(2)}$$

The second rule states that the pressure produced by the fan varies as the cube of the fan speed. This is also expressed as a ratio between fans:

$$\frac{P(2)}{P(1)} = \frac{\left(RPM(2)\right)^2}{\left(RPM(1)\right)^2}$$

The third rule states that the horsepower required to drive an air moving device varies as the cube of fan speed. This is expressed as follows:

$$\frac{HP(2)}{HP(1)} = \frac{\left(RPM(2)\right)^3}{\left(RPM(1)\right)^3}$$

In centrifugal loads, even a minor change in the motor's speed translates into a significant change in energy consumption. Energy conservation benefits can be achieved with energy efficient motor retrofits.

HARMONIC DISTORTION

Harmonics may be introduced into the electrical system due to the nonlinear characteristics of some power loads and their operation. Harmonic components of current and voltage are multiples of the normal 60-Hz AC sine wave. These harmonics can cause motor, transformer and conductor overheating, capacitor failure, malfunction of relays and controls and reduce system efficiencies. These problems can be reduced by adhering to IEEE-519 Recommended Practices and Requirements for Harmonic Control in Electrical Power Systems.

k-FACTOR TRANSFORMERS

Harmonic producing loads are raising the temperature of transformers to unacceptable levels, particularly in those that are close to the connected nonlinear loads. A newer type of transformer is designed to cope with the effects of harmonics. These are transformers with k ratings. Prior to the following UL standards:

Standard 1561   Dry-type General Purpose and Power Transformers
Standard 1562   Transformers, Distribution, Dry-type—Over 600 Volts

The loading of a transformer was restricted to a sinusoidal, nonharmonic load with less than 5% harmonic content. The above standards allowed UL to list specially constructed transformers for powering harmonic-generating loads. These transformers have the following notation on their nameplates:

Suitable for Non-Sinusoidal Current Load
with a k-Factor not to exceed xxx.
The k rating attempts to cover the following
nonlinear load effects on transformers.

Harmonic currents may cause the overheating of conductors and insulating materials due to a phenomenon known as the skin effect. At lower frequencies such as 60-Hz, sinusoidal current will then flow through the entire cross section of a conductor. This provides the lowest conductor resistance and loss. As the frequency increases, current tends to flow more toward the conductor's surface and away from its cross section. The conductor resistance and loss can increase substantially. In the case of phase-to-neutral nonlinear loading and resultant third harmonic content, the 180-Hz current on the neutral conductor will exhibit this property.

When direct current is passed through transformer windings, a simple I^2R loss results, where R is the DC resistance of the winding. When the same magnitude of alternating current goes through these same conductors, an additional stray loss is produced.

The portion in the transformer windings is called eddy-current loss. The portion outside the windings is called stray losses. The eddy-current loss is proportional to the square of the load current and the square of the frequency so nonsinusoidal load currents can cause very high levels of winding loss, which also cause a winding temperature rise.

Core flux density increase is another problem. Transformers, and motors are usually the largest ferromagnetic loads connected to the distribution system. The load characteristics of motors tend to be more linear than transformers since motors have an airgap. Power transformers may suffer from partial saturation of the transformer's iron core which will cause harmonic magnetizing currents that differ greatly from a sinusoidal waveform.

Triplen harmonic currents like the 3rd, 9th and 15th can circulate in the primary winding of a delta-wye transformer. They add to the core flux density and may drive it towards saturation. The 5th, 7th, 11th and 13th... harmonic currents will pass directly through a delta-wye transformer to its source of power and can cause overheating and damage to the source.

Before k-rated transformers were available, the only way to protect these transformers was to limit the connected nonlinear load to some value less than the transformer's full load rating (derating the transformer). This load derating might range from 20 to 50%. The Computer & Business Equipment Manufacturers Association (CBEMA) recommends a derating calculation for transformers feeding phase-to-neutral nonlinear loads with a high 3rd harmonic content and crest factor.

ANSI/IEEE C57.110-1986 Recommended Practice for Establishing Transformer Capability When Supplying Nonsinusoidal Load Currents, provides derating calculations for distribution transformers powering phase-to-phase and phase-to-neutral nonlinear loads. The method derates a standard, nonrated transformer by calculating the additional heating that will occur for a load that generates specified amounts of harmonic currents. This derating of the transformer allows it to handle the additional heating without damaging itself or reducing its useful life.

Using a standard transformer to power a nonlinear harmonic rich load, even when derated, violates its Underwriters Laboratories (UL) listing. Standard transformers are not rated or tested for use with significant nonlinear loads.

k FACTOR CALCULATION

ANSI/IEEE C57.110-1986 provides a method of calculating the additional heating, (wattage loss) that will occur in a transformer when supplying a load that generates a specific level of harmonic distortion. The method involves calculating the per unit RMS current for each of the harmonic frequencies. When the individual and total frequency harmonic distortions are obtained from a harmonic analyzer, the individual frequency per unit RMS current can be found from the following expression:

$$\%HD(h)/\sqrt{(THD)^2 + (100)^2}$$

Here:

$$
\begin{aligned}
h &= \text{harmonic frequency order number } (1, 3, 5,\ldots) \\
h\%HD(h) &= \text{percent harmonic distortion at harmonic order } h \\
THD &= \text{total harmonic distortion.}
\end{aligned}
$$

C57.110-1986 also shows how to calculate the additional transformer wattage loss as a result of a specific distribution of harmonic currents. Each RMS current calculated above is squared, and then multiplied by the square of the harmonic. This is done for each of the harmonic orders present and then added together. This summation is called the k factor.

It is multiplied by the eddy-current losses, which provides the increased transformer heating due to the harmonic distortion. Eddy-current losses are found by subtracting the I^2R losses from the measured impedance losses.

k factor is a function of both the square of the harmonic and the square of the harmonic current. This calculation procedure is best done using a computer spreadsheet. The following steps are used for the calculation.

1. List the harmonic order number (h) for each harmonic generated, including the fundamental or first harmonic (60-Hz).

2. Square each harmonic number.

3. List the harmonic distortion values for each harmonic (including the fundamental) as obtained from a harmonic system analyzer. Enter them as decimal equivalents (30% = .30).

4. Square the harmonic distortions harmonic including the fundamental. Then, add all the squared distortions.

5. The square of each distortion value is multiplied by the square of its harmonic number.

6. Add all these values.

7. Divide this sum by the sum of the squared distortion to obtain the k factor for that piece of equipment.

Typical k-rated transformers are 600-V ventilated dry types with k ratings of 4, 13, 20, 30, or 50. They usually have delta-connected primaries and wye-connected secondaries. Temperature rises of 150, 115 and 80°C at rated voltage are available. Most use a 220°C UL recognized insulation system. Windings can be copper or aluminum. Units are available with six 2.5%, two FCAN and four FCBN. Most have electrostatic shields between the primary and secondary windings to limit electrical noise.

To compensate for skin effect and eddy-current losses, the primary and secondary windings use smaller than normal, paralleled, individually insulated, and transposed conductors. For the third harmonic currents that circulate in the primary delta winding, this winding is sized to limit the temperature rise to the transformer's rated value. The secondary neutral is sized to have twice the capacity of the secondary phase conductors. The basic impulse level (BIL) is normally rated at 10kV.

Harmonic currents generated by modern electronic loads will remain until manufacturers include filtering equipment in their products to eliminate them. Until then, the k-factor will be a consideration in electrical distribution systems.

The k factor calculation makes some assumptions. All of the stray loss in the transformer is assumed to be eddy-current loss. The I^2R loss is assumed to be uniformly distributed in each transformer winding. The eddy-current loss between the windings is assumed to be 60% in the inner winding and 40% in the outer winding for transformers with a maximum self-cooled rating of less than 1000A, regardless of turns ratio. For transformers having a turn ratio of 4 to 1 or less, the same percentage split is used. For transformers with a turns ratios greater than 4 to 1, and with a self-cooled rating greater than 1000A, the loss is 70% to the inner winding and 30% outer winding. The eddy-current loss within each winding is assumed to be nonuniform, with the maximum eddy-current loss density assumed to be near the transformer hot spot. This maximum density is assumed to be 400% of the average eddy-current loss in that winding.

Higher order harmonics will generate more heat per unit of current than lower order harmonics. Higher order harmonics mean higher k numbers.

In most electrical distribution systems, the highest levels of harmonic distortion occur at the receptacle or equipment connection point. The harmonic level at the panelboard is much lower. Levels at the service transformer are quite low, and a k-rated transformer is not usually required at this point.

Phase shift is an important factor. The phase relationship is used by certain equipment manufacturers to cancel harmonics in electrical distribution systems.

Harmonic cancellation and source impedance effects explain why a total load k-factor greater than 9 does not usually occur in actual office electronic loads or any similar concentration of switched-mode power supplies.

The total k-factor of office load systems does not correlate with the relatively high values of harmonics seen at individual branch load circuits. Most commercial office loads operating at or near the feeder transformer's full load rating do not exceed k-9.

A harmonic analysis of an office with a single personal computer and video monitor will typically measure a k of 14 to 20 in harmonic intensity. For an office filled with large number of such PCs, a transformer with a k of 20 or 30 is needed to feed the distribution panel serving the office power. These offices usually contain other equipment, such as copiers, printers and FAX machines. The k-factor for all the equipment should be listed, weighted by their relative current contribution and averaged for a total k-factor estimate.

The k-factors of electronic loads in parallel do not simply average out. Fifth and higher-order harmonics in the load current from multiple single-phase electronic loads on a given feeder occur at different phase angles, which change during operation. This distribution of harmonic phase angles results in a reduction or cancellation of higher frequency harmonics.

As more current flows in a feeder, the source impedance of the supplied power begins more of an effect, especially for the higher frequency harmonic components in the current. If 20 devices are on-line simultaneously, the combined k-factor at the bus of the distribution panel is reduced by a factor of three or more below the average individual device k-factor. In general, the higher the number of single-phase nonlinear loads on a given distribution panel, the lower the k-factor.

Harmonic measurements show a distinct reduction of harmonics as loads are added. Reduction of the fifth and higher harmonics is thought to be primarily due to harmonic cancellation. Other reduction effects are attributed to feeder impedance.

In one test, the k-factor dropped from the initial 13.9 with just a few devices to 4.6 with 26 devices on-line. The third harmonic was reduced from about 90% of the fundamental to less than 65% as the load current increased. Since the third harmonic does not cancel, this reduction is mainly due to source impedance. Since the transformer feeding the office loads is usually the major ohmic impedance in the feeder, the reactance of the transformer plays a major role in controlling neutral current caused by the third harmonic.

THE h FACTOR

Today, a rapidly growing number of the receptacle-connected loads are nonlinear, especially those found in commercial office spaces with many personal computers and other office equipment. In a typical commercial office space, a distribution transformer might serve several hundred circuits. These circuits could be connected to many nonlinear loads. When selecting a transformer for an existing installation, load currents can be measured at the point where the transformer is to be installed and the k factor calculated. In a new system however, the conditions and relationships are not known.

Over a period of years, case studies of the k-factor have been documented, mostly from commercial office buildings and data centers. This information has been used to provide a set of criteria to determine k-factors.

Another parameter, called the h-factor can be used. The h-factor is used to quantify the harmonics created by the load and refers to the heating effect or impact of a nonlinear load on transformers and similar components. The h-factor refers to the causes of the harmonics.

Wiring, power source impedance, and phase angle can limit or reduce the h-factor. The h-value is usually less at the circuit's source (breaker panel or transformer secondary) than at the load connection point (receptacle). The h factor at the panelboard feeder will generally be lower than the lowest individual h factor, if a significant linear load is not present.

To use the h factor you must determine the h value of the present load on the transformer by assigning an h factor for each existing load. The equipment manufacturer may provide this information, or you can measure the h factor for an existing piece of equipment. PCs, monitors, terminals, printers and other office equipment represent a typical load for many office buildings. The average value of the h factor for these buildings is 9.6. This means that a transformer with a k rating of 10 or greater would satisfy the requirements of most of these installations. A more conservative value would allow for some expansion. This would mean a transformer with a minimum k rating of 13 or more. Depending on the actual load of the transformer, it could be necessary to reduce the load to allow operating the transformer at its rated temperature or to change the transformer to one with a higher k rating.

The h-factors of loads (linear and nonlinear) cannot be added arithmetically and the impedance of the feeder and branch circuit wiring limits the effects of the harmonic currents produced by the nonlinear load. However, the phase relationship of individual harmonic currents and individual load currents provide some cancellation and reduction of harmonic distortion.

If the load is linear, only one magnitude and angle are present but with nonlinear loads, each load is represented by magnitudes and angles for the fundamental and each of the harmonics.

If the h-factors for the loads are known, they must be referenced to a common impedance. Both the point of measurement and the impedance of the circuit will affect the h-factor of a load. When comparing loads connected to a branch circuit, the h-factor should always be measured at the load terminals on a branch circuit with the same length and type of wire and conduit. The current measured at the load end of a branch circuit serving a nonlinear load is higher than that measured at the circuit breaker.

Individual nonlinear loads, such as the personal computer, monitors and other equipment produce high levels of harmonics with h-factors of 10 to 30 or more. This is true of all phase-to-neutral connected electronic equipment. Phase-to-phase connected equipment can also produce similar h-factors, depending on their design and the way they are operated.

The use of variable frequency drives (VFDs) is growing as the need for energy efficient design increases. These drives have a high h value. These drives should be isolated from the electrical system when possible by using a transformer with a suitable k rating or by installing series line reactors ahead of the VFDs.

The electrical wiring will attenuate, or diminish, harmonic currents. The higher the harmonic number, the greater the attenuation. Source impedance also affects the harmonic distortion, with higher impedance reducing harmonic currents. Too high an impedance results in excessive voltage distortion and poor voltage regulation.

Partial cancellation of some harmonics will result from widely varying phase angles. Thus, the h factor at the service panel, even with high h factor nonlinear loads, will be lower than that of each individual nonlinear load.

OVERSIZING TRANSFORMERS

One 208Y/120V office system, consisting of computer and printer loads, was supplied by a 45kVA transformer that was only about 35% loaded. The transformer was temporarily replaced with a 225kVA unit to observe the effect of lower ohmic resistance on the load harmonics. The transformer substitution caused the k-factor to increase from the normal value of 4.75 to a much higher value of 8. The neutral, which was already at 120% of average line current, jumped to over 160%. This shows that oversizing of supply transformers is undesirable and that the existing 45kVA supply transformer should be replaced with a smaller unit to reduce the neutral current. Transformers run most efficiently at 70 to 80% of their capacity.

Oversizing of transformers or using an unnecessary high k-factor rated transformer can increase neutral conductor currents along with damaging neutral-to-ground voltages at sensitive electronic loads. Only at or near transformer full load is a k-rating necessary to prevent overheating of transformers. Since transformer losses vary with the square of current, the reduction or loss at reduced loading has to be considered with the offsetting heat buildup from an increased k factor.

At full load, the transformer reactance has the most effect in reducing load harmonics. The Computer Business Equipment Manufacturers Association (CBEMA) recommends transformer impedances of 3 to 5%. This provides some reactance to prevent excessive harmonic current flow and resulting high neutral currents, but high enough to cause excessive voltage distortion. Transformers with a k of 20 or 30 for office applications are based on data from individual branch loads and cannot be justified by the total load k-factor.

In most office applications, k-9 or k-13 transformers are a good choice for high office loads. As more users measure the actual levels of k-factor in their systems at near full load, the k-4 transformer rating may even become popular.

Specifying a k-factor of more than 13 can create problems from low impedance, a typical symptom of oversized transformers. Many k-20 and k-30 transformers are below the 3% minimum impedance recommended for computer loads. Many, also have very low X/R ratios, lowering the reactive component even further.

POWER SYSTEM MEASUREMENTS

The measurement of today's power loads involves more than kilowatts. The use of motor drives, electronic ballasts, power conditioners and other loads requires a more complex evaluation. In the past, adequate evaluations could be done with analog line voltage power meters and multimeters.

Today, measurements of true power factor, displacement power factor, true-rms current, and total harmonic distortion are needed. These measurements can be done with an advanced line voltage power meter.

Traditional power meters use analog multipliers or Hall-effect devices to measure power consumption. They make accurate measurements, but the analog circuits cannot measure harmonic distortion. Newer power meters usually have a sampling circuit and a microprocessor.

These power meters have enhanced memory capabilities and comprehensive recording abilities. A data interface, usually a serial port, allows the downloading of data to a personal computer, where it can use other applications for reports, analysis and trending. Time profiles can be stored on CDs or other media for archive use. Data can be sent to a Web site for collaboration, reporting, and searching.

Measuring involves making direct connections to line power, often at the switchgear or feeder level. The continuous voltage rating of the meter should be higher than the worst case measurement requirement. A 600V rating would be used for evaluating a load connected to a 480V source.

The IEC 1010 Overvoltage Category defines how much transient overvoltage a meter can withstand without damage. IEC1010 establishes

Overvoltage Installation Categories 1-4. The higher overvoltage category number refers to an electrical environment with higher energy available and the potential for higher voltage transients. A category 2 rated instrument should only be used at the end of a long branch circuit. A category 3 rated instrument should be used on a load connected directly to a service panel.

Inductive current measurements are preferable to methods that require breaking the circuit. Current clamps have upper and lower limits for their accurate measuring range. Exceeding the upper limit is not just a matter of accuracy, it is also important for safety. To measure below the clamp's lower limit, wrap several conductor turns through the jaws. Make sure the jaws close properly and make your measurement. Then, divide the measured value by the number of turns, to get the true measurement.

A major lighting retrofit is a good time to improve power quality. A test of a few ballasts from each manufacturer will show the actual performance. The tests should include the power in watts, displacement power factor, true power factor and the total harmonic distortion (THD) of the line current.

When THD is expressed as the percentage of harmonic current with respect to the fundamental, it is called the THDF. If it is expressed with respect to the rms current, it is called the TDHR. A power meter shows both values and allow a comparison with the manufacturers data which may list THDF or THDR specifications.

Electrical performance is only one factor in a lighting upgrade. There is the light output, expected lifetimes, radiated EMI, fixture costs and the cost of installation. You can also evaluate motor drives and office machines with similar test criteria.

DISPLACEMENT POWER FACTOR

When loads are linear, such as conventional induction motors, the voltage and current are essentially sine waves, and a form of power factor called the displacement power factor (DPF) is present. Displacement power factor indicates where the load inductance shifts or displaces the current from the voltage. True power factor will account for the effects of harmonic current. True power factor is always lower than displacement power factor when the line current contains harmonics.

The ratio of useful working current to total current in the line, or the ratio of real power to apparent power, equals the DPF. The use of inductive equipment such as induction motors, requires the power lines to carry more current than is actually needed to perform a specific job. Capacitors can supply a sufficient amount of leading current to cancel out the lagging current required by the electrical system, improving system efficiency.

Today, many electrical system also have harmonic currents on their lines. Harmonic currents caused by nonlinear or pulsed loads, such as electronic power supplies cause the apparent power to exceed the active power.

In these situations, the form of power factor present is called distortion power factor. The sum of the displacement and distortion power factors is the total power factor (TPF).

For linear loads, measurements can be made of the displacement power factor with hand-held instruments. These instruments measure kilowatts (kW) and kilo-volt-amperes (kVA). Some can directly read power factor (PF). When harmonics are present, meters with true rms capability are required to measure the total current which is the current at the fundamental 60-Hz and the harmonic currents to determine the TPF. When the total current, including all the harmonic currents on the line, is used in determining the apparent power (kVA) and the active power (kW), then the TPF equals the rms values of kW divided by kVA.

Both power factor values are critical since power factor is a measure of how efficiently a facility uses the capacity of its power systems. They determine how much energy, both work-producing and non-work-producing, is required to power a load. Work-producing energy is measured in watts, nonwork-producing energy is measured in VARs.

Apparent power is the total system capacity needed to perform a certain amount of work. The kVA equals volts times amps, each expressed as root-mean-square (RMS) values. As the amount of kVARs (nonwork-producing energy) increases, so does the inefficiency of the electrical system.

Harmonics increase the apparent power required to do a particular amount of work. When active power is divided by apparent power in the presence of harmonics, the result is the Total Power Factor (TPF or PF). The component not contributed by harmonics is the displacement power factor (DPF).

TPF and DPF differ in any circuit with nonlinear electrical loads because these loads create harmonics. They are the same in completely linear circuits, even if phase shifts between voltage and current occur because of inductive or capacitive loads.

The harmonic currents caused by nonlinear loads may cause TPF to be low (.60 to .70) while the DPF could be relatively high (.90 to .95). Because of the abundance of nonlinear loads now being used, the actual PF should be considered as the total power factor.

In the past, when the electrical system had a low PF, the usual procedure was to add capacitors without reactors, which was done to avoid tuned circuits and to reduce harmonics. That can still be done when most of the loads have a 60-Hz sinusoidal signature. Today, adding pure capacitance to correct a low PF situation may cause problems due to the harmonics in the electrical system.

The impedance of capacitors decreases as the system frequency increases, and harmonic currents are multiples of the fundamental 60-Hz current, so capacitors become sinks that attract high-frequency currents, causing possible overheating and early failure. The answer is to install a filter (capacitors, reactors, and resistors) that trap the harmonics. These will improve the power factor but the use of reactors and resistors will reduce the flow of harmonic currents.

Capacitance and inductance in an electrical system may form a tuned circuit where the current is resonating at a specific frequency. This is the frequency where the capacitive reactance equals the inductive reactance. If the circuit is exposed to an exciting harmonic, at or close to the resonating frequency with sufficient amplitude, the current in the circuit will oscillate, with a circulating current.

The resulting circulating current can produce a high voltage across the circuit elements and the high current flow can blow fuses, damage components and cause an excess of harmonics in the electrical system. Harmonic currents affect motors and transformers and can cause excessive heating.

The resonant frequency can be shifted by adding or removing capacitors from the capacitor bank or by relocating the capacitor bank to change the amount of source inductance that is available to the capacitance. Moving a PF correcting capacitor away from a harmonic producing device adds some damping that reduces the harmonics. For power distribution systems with harmonics, where several motors are connected to a bus, the capacitors should be connected to the bus.

The capacitor bank can be also be modified into a filter bank by installing the appropriate reactors and resistors. If the resonant frequency cannot be changed, filters can be added to reduce the harmonics. These additional filters need to be analyzed so they do not cause resonance at a lower frequency.

Two types of filters are used. A shunt filter across the load will appear as a short or low impedance at 60-Hz. A filter in series works in the opposite manner. It provides a high impedance to harmonic frequencies and a low impedance to 60-Hz. A disadvantage of series filters is that they must be rated for the full load current in the line along with the harmonic currents.

There are several other methods to reduce harmonics. If the resonant inductance is a motor winding, changing the motor speed by changing the number of poles is a possibility. The correct use of 3-phase transformers can also reduce harmonics. A delta-wye transformer will tend to trap the 3rd, 9th, 15th and the other triplen harmonics. A grounded wye-grounded wye transformer does not. The 5th, 7th, 11th, 13th and other harmonics other than the triplen frequencies will pass through any type of transformer, but they will be attenuated by the transformer impedance.

HARMONICS AND POWER READINGS

The difference between PF and DPF readings is proportional to the amount of harmonics in the power distribution system. If only a single power factor is specified for a piece of equipment, the specification usually refers to PF.

When a linear load like an induction motor is measured, PF and DPF are equal. Since the motor is a linear load, there is a phase-shift due to inductance and DPF represents the total power factor.

Three-phase full-wave rectifiers are typical of three-phase nonlinear loads, others are static power converters and adjustable-frequency drives.

They have a DPF somewhat higher than the total PF.

Due to the pulsed nature of the current demanded by nonlinear loads, these loads have a high harmonic content. This current can be a significant part of the true-RMS current.

If PF and DPF differ by 10% or more, harmonics is usually the cause. In a predominantly linear system PF and DPF are essentially the same value and motors or other linear, inductive loads generally dominate. In this system, a low power factor can be compensated for with correction capacitance. Even if PF and DPF are identical, readings under 0.90 indicate that system modifications may be required. But, improperly applied capacitors can cause resonant conditions that can lead to over-voltages even in systems with low levels of harmonics.

In predominate nonlinear systems, where PF and DPF differ significantly, and the PF is lower, the safest way to correct low power factor generally is to reduce the harmonic currents. This requires locating the harmonic current sources and applying line reactors or filters. Using capacitor networks with series inductors will limit the harmonic current in the capacitors, but correction capacitors and compensating filters must be used with caution to prevent resonance problems at harmonic frequencies.

Testing of a facility's electrical system to verify existing harmonics, PF and 24-hour load profiles should be done when planning low PF and harmonic solutions. In systems with capacitors already installed, if adjustable-frequency drives are added, the new components may cause instability and overvoltages. Readings should be taken in the circuit to determine if the correction capacitors need to be removed.

Readings should be made with a power harmonics meter. Some newer meters provide a readout of the signal parameters along with a visual display of the waveform and a view of the entire harmonic spectrum.

The waveform display shows the effects of the harmonics. The more distorted the sinewave is, the worse the harmonics. The harmonic spectrum shows the presence of individual harmonic frequencies and indicates the source of the problem.

ADJUSTABLE CAPACITOR BANKS

Capacitor banks can be designed to correct low PF at different kVA load levels to match the facility's electrical load. The capacitor bank can be split into several steps of PF correction. Automatic controls are used to change the switching devices. Resonant conditions should be checked using harmonic analysis software before the filter bank is employed. Different combinations of filters are needed to dissipate specific harmonics.

The normal procedure is to switch in the lower order filters first, and then add the higher order filters. The procedure is reversed when filters are removed from service. This is done to prevent parallel resonant conditions that can amplify lower frequency harmonics. These conditions can be caused by the higher frequency filters.

POWER FACTOR AND HARMONICS STUDIES

North Star Steel of Beaumont, Texas, conducted a study for low power factor and harmonics problems before it installed a new ladle furnace. The study consisted of computer simulations and field measurements. The objective was to raise the plant's power factor to above 0.90 lagging which would eliminate utility penalties. The study showed the need for harmonic filters to compensate for the reactive load of the new furnace and correct existing problems.

The plant had two existing scrap metal furnaces, and the electrical system included corrective PF capacitors as components of the existing harmonic filter banks. A rolling mill connected to a 13.8-kV system with no PF correction was also in operation.

Steel mills with arc furnace and rolling mill loads are particularly subject to harmonics and PF problems. These loads operate at PFs that may result in penalties and lower bus voltages. The nonlinear characteristics of furnace arcs and rolling mill drives may generate significant harmonic currents in a plant's electrical system.

Harmonic currents are produced at scrap metal or ladle furnaces from the harmonic voltages of the arc impressed across the electrodes. This load current is passed through the impedance of the cabling and furnace transformer. These harmonic currents are injected back into the electrical system and may cause problems if the system is resonant at one of the predominant harmonic frequencies. The harmonic currents will excite the resonant circuit, producing high rms and peak voltages. This may cause equipment damage or degradation. The voltage distortion can disturb electronic power equipment such as drive systems and may interfere with the operation of control systems.

The harmonics produced by scrap metal and ladle furnaces vary due to the changing arc length over the total heating period. Scrap metal furnaces predominantly generate a 3rd harmonic voltage and produce an erratic voltage total harmonic distortion (THD) on the bus. Ladle furnaces predominantly generate 3rd and 5th harmonic voltages and produce a more consistent THD. The erratic arcing behavior also produces an unequal current conduction in the positive and negative half cycles.

An arc furnace load can be represented as a harmonic voltage source in series with some load impedance. This is the impedance of the cables and electrodes in the furnace.

A steady-state harmonic model of the plant was developed based on the existing loads, conductors, and field measurements. Testing was conducted at different locations in the facility to determine the existing harmonic voltages and currents resulting from scrap metal furnace and rolling mill operations.

Measurements at the primaries of the furnace transformers, the main utility entrance busses, the existing harmonic, filter banks, the main feeders serving the 13.8kV bus, and the primary feeders for the rolling mill were made. Testing was carried out during operation of the scrap metal furnaces (arc load) and the rolling mill (drive load). For the future ladle furnace, data obtained from harmonic field measurements of similar furnace installations was used.

Using portable harmonic monitoring instruments, on-site measurements were performed at the main switch gear feeder circuit breaker locations on the 13.8kV system for the rolling mill loads over a period of several days. Existing current and voltage transformers were used as signal sources. Power flow and power factor data also were gathered.

The scrap metal and ladle furnace harmonic voltages are shown in Table 4-1.

The voltage depends on the load levels of the mill stands that mill or roll the steel into various shapes and thickness. When the mill stands are worked hard, there is a high current flow and high THD. When the machines are idling between loads, there is a low current flow and low THD. The current THD was in the range of 3.4% to 10% of the funda-

Table 4-1. Measured Harmonics - Steel Mill

Harmonic	Scrap Furnace	Ladle Furnace
2nd	5.0%	2.0%
3rd	20.0%	10.0%
4th	3.0%	2.0%
5th	10.0%	10.0%
6th	1.5%	1.5%
7th	6.0%	6.0%
8th	1.0%	1.0%
9th	3.0%	3.0%
11th	2.0%	2.0%
13th	1.0%	1.0%

mental current. The total power flows into the 13.8kV bus were measured.

The mill stand motor drives produced the largest harmonic content. They are connected through step-down transformers and the 6-pulse units generated a dominant 5th harmonic current.

Power for the rolling mill is divided into two main circuits. One circuit feeds the 6-pulse drives. The total harmonic distortion (THD) on this circuit was about 30% of the fundamental current. It contained 5th, 7th, 11th, and 13th harmonics. The THD measurement on the other circuit was less than 8% of the fundamental current.

Harmonic field measurements for the 34.5-kV system was normally less than 7%, but during erratic furnace arcing conditions, voltage THD values increased to almost 20%. The current THD on the two main feeders ranged from 3% to 16% of the fundamental current. The scrap metal furnaces are a low PF load, typically 0.65 to 0.85 lagging. With two harmonic filter banks already operating, the overall PF ranged from 0.90 lagging to 0.90 leading.

Reactive compensation was needed at the main 13.8kV bus to correct the PF from an average value of 0.76 lagging to 0.95 lagging. The existing harmonic filter banks on the 34.5kV system can provide ample compensation for the additional reactive load of the new ladle furnace.

This is because the 34.5kV system, at times, operated at 0.90 leading PF. The existing capacitors of the harmonic filters had the capability to correct the low PF from the new load as well as serve the existing load.

Measurements indicated the 13.8kV system load was approximately 23-MW with a 0.76 lagging PF, with an uncorrected reactive load of 20Mvar. To improve the 13.8kV system PF to 0.95 lagging, the reactive load had to be reduced to 7Mvar. An additional capacitor bank of 13Mvar can provide the reactive compensation.

HARMONIC FILTERS

A harmonic computer program can perform a steady-state analysis of the facility's electrical system for each frequency at which a harmonic source is present. The program will calculate the harmonic voltages and currents in the system. In these harmonic simulations, PF correction capacitors can be connected to check for system resonance.

A harmonic computer simulation of the 13.8kV power system, with 13Mvar reactive compensation added to the main bus, was made and revealed a parallel resonant peak close to the 5th harmonic.

The computerized harmonic analysis indicated that the 13Mvar capacitor bank tripled the voltage THD and the current THD reached values as high as 62% due to the parallel resonant peak. The parallel resonant condition near the 5th harmonic caused harmonic amplification. A filter was needed at this frequency. An in-line reactor was needed to convert the capacitor bank into a harmonic filter. The required equations are:

$$h^2 = X_C/X_L > X_L = X_C/h^2$$

$$X_C = (kV_{L-L}^2)/Mvar_{3-phase}$$

Where:

$$h \ = \ \text{Tuned Harmonic}$$
$$X_C \ = \ \text{Capacitive Reactance of Filter}$$
$$X_L \ = \ \text{Inductive Reactance of Filter}$$
$$kV_{L-L} \ = \ \text{Rated Voltage of Capacitor Bank}$$
$$Mvar_{3-phase} \ = \ \text{Rated Mvar of Capacitor Bank}$$

In the model of the electrical system, the capacitor bank was changed into a harmonic filter bank by adding reactors and resistors. This resulted in a decrease in voltage and current THD from the values obtained during testing prior to the installation of the filter bank. The maximum voltage THD was below 3% and the maximum current THD was below 4%.

Capacitors used with reactors as part of a filter will experience a steady-state voltage above the nominal line-to-neutral voltage of the electrical system. This excess voltage is a function of the harmonic at which the filter is tuned. When installing capacitors in a harmonic filter, the voltages and currents should be within the following limits established by ANSI standards regarding capacitor nameplate ratings:

RMS voltage < 110%
Peak Voltage < 120%
RMS current < 180%
Reactive Power Output < 135%

Harmonic analysis can be used to check these values at the filter bank capacitors.

TRANSFORMER INRUSH CURRENT

Another harmonic current source is the inrush current from transformer energizing. This includes even and odd harmonics that decay with time until the transformer magnetizing current reaches a steady-state condition. Along with the fundamental current, the most predominant harmonics during transformer energization are the 2nd, 3rd, 4th, and 5th. These harmonics normally do not cause problems unless the system is resonant at one of these predominate harmonic frequencies produced by the inrush current. Then, transformer energization can excite the system, causing voltage distortion that can impact the energizing inrush current. This can produce more harmonic currents and cause further distortion. The high values of rms and peak voltages can degrade or damage equipment and lead to premature equipment failures.

CAPACITOR EFFECTS

Capacitors can sink a fair amount of harmonics, but, they may interact with large harmonic producing loads like arc furnaces or rolling mills. This can result in magnification of the voltage and current distortion. The increase in the distortion is usually due to a parallel resonant condition. This condition exists when the electrical source's impedance (inductive reactance) is equal to the capacitor bank's impedance (capacitive reactance) at a common frequency. It results in a tuned circuit with a circulating current.

One indication of excessive harmonics at a capacitor bank is an increase in blown fuses. This can result in eventual failure of capacitors. When fuses blow in a capacitor bank, the parallel resonant frequency shifts to a higher frequency, sometimes resulting in a stable operating condition. When the blown fuses are replaced, problems may reoccur, since this returns the system to the original parallel resonant frequency that caused the initial fuse failure.

LOAD INTERACTIONS

Transient problems which can disturb or destroy equipment includes lightning disturbances which are among the most recognizable of these events. But, load switching surges and the transients associated with electrical current inrush can also be disruptive. One such transient comes from the Silicon Controlled Rectifier (SCR) assembly used in the rectifier section of a variable speed drive device.

Many power conditioning problems are interactions with load devices. A solid-state Uninterruptible Power Source (UPS) may not synchronize with the load it is carrying and passes the load off to the bypass line. A standby generator cannot be used to power even an unloaded UPS, because the current demands of the UPS rectifier are so distorted that the generator must be several times the size of the UPS to provide the correct energy. These problems can often be explained by harmonic distortion demands.

Most loads are designed to operate with 60-Hz electrical power, but under some conditions the incoming service, transformers, busways, panel boards, branch circuits, and load distribution devices may demand several frequencies.

An UPS may be installed in order to keep all loads running even when the power company has an anomaly in their service. The UPS supplies 60-Hz power, which is supported by batteries for the time needed. However, suppose the loads also need 180-Hz, 300-Hz, 420-Hz, and other frequencies.

Newer loads, for both single phase and three phase equipment, may make demands upon the power source for a mixture of frequencies for the current they take to operate. The higher frequencies above 60-Hz are a part of the current spectrum which characterizes these non-linear devices. These extra frequencies cause heating in the wiring and electrical distribution equipment, lowering the efficiency of the system in handling energy. In a three phase load, these extra currents may reduce the power factor by 10 to 15%. This creates an unwanted peak energy demand for the system. If enough of these devices, such as variable speed drives (VSDs) are installed on a system, the resultant harmonic currents could be large enough to cause voltage distortion in excess of the amount which could be handled by the power supplier.

In single phase loads, such as those caused by personal computers or electronic lighting ballasts, the loss of capacity starts on the common neutral wire of three phase, four wire distribution systems. The overloading of the neutral occurs due to the high content of 3rd harmonic, 180-Hz, which does not cancel when each of the phase contributions arrives on the common neutral.

A 750-kVA distribution transformer, 480 volts to 208/120 volts, may only carry 100 amperes on each of the secondary phase wires when 1/2 loaded. But, harmonic currents can increase the neutral current into the transformer to over 200 amperes. The current in the neutral should cancel, except for the unbalanced portion at 60-Hz.

New drive equipment often lowers the power factor. But, adding power factor correction capacitors to improve the power factor and avoid penalty billing may cause the circuit breaker to trip from the inrush current to turn on the capacitors.

High frequency currents from an elevator controller caused such a voltage distortion that 120 volt control circuits in energy management devices were damaged. The manufacturer of the elevator controller needed to add filters for the distorted currents.

MOTOR REPLACEMENTS

Replacing a standard motor with an energy efficient motor in centrifugal pump or a fan application can result in increased energy consumption if the energy efficient motor operates at a higher RPM. Even a small RPM increase in the motor's rotational speed can negate any savings associated with a high efficiency motor retrofit.

An investment of 20 to 25% for an energy efficient motor, over the cost of a standard motor, can be recovered in a short period of time. In some cases the cost of wasted energy exceeds the original motor price in the first year of operation. Performance evaluations of all motors should be done routinely. Motor maintenance will keep the building or plant running smoothly with less downtime due to failures.

POWER PROBLEMS

Most equipment requires continuous 120-volt AC power (plus or minus 10 volts). Except for occasional outages, this is what the utility company usually delivers. Lightning, floods, storms, accidents and large loads coming on-line at once can corrupt commercial power.

Power problems are more common than most people realize. A study performed in the early 1970s by IBM and Bell Laboratories concluded that power problems sufficient to interrupt computer operations occur an average of twice a week. Power outages, or blackouts, account for less than 5% of these twice weekly power problems. Other, less serious types of commercial power degradation occur every day. Left unchecked, these power imperfections can damage sensitive equipment.

Disturbances to commercial power include the following anomalies:

- spikes—A spike is a burst of high voltage lasting for only a fraction of a second. Spikes are also known as impulses or transients.

- noise—Noise consists of high-frequency, low-voltage oscillations or ripples that originate from sources such as fluorescent lights and heavy equipment motors.

- surges—A power surge is a sudden increase in voltage that can last several seconds.

- sags—A sag is a decrease in voltage lasting up to several seconds. Sags are a large majority of power problems.

- brownouts—A brownout is a prolonged sag that usually occurs during periods of peak power usage, when the local utility company's power producing capacity is severely drained. A typical example is a hot summer day when air conditioners are on.

- blackouts—These are also called a power failure or outage, a blackout is the total loss of power. Blackouts are relatively infrequent.

Blackouts are infrequent, but their effects are potentially the most devastating on facility operations. If the blackout occurs during a critical time, important data and other material loss is imminent.

The effects of a blackout are immediately obvious, but the effects of other types of power disturbances may not be so obvious nor predictable. Power disturbances can corrupt some of the data held in computer memory. If the disturbance occurs during a particular system routine, the system may continue to operate normally until that routine is used again. Then the system may lock up or crash because of the corrupted data in memory.

Power surges are a common cause of equipment damage. Surges can stress electronic components and cause them to fail. High frequency noise or spikes can cause noise, read/write or data parity errors. All of these can cause computer communications problems. Spikes can also cause component damage.

Power disturbances may occur in combination, which compounds their effects on the equipment. Slight sags in power can occur frequently and may not be sufficient to adversely affect equipment. But, if these sags occur during a brownout, the power level may be reduced below the acceptable power range that the equipment needs to operate.

POWER PROTECTION DEVICES

A number of power protection devices are available, but not all of these devices protect against the full range of power problems. Some, such as surge suppressors and voltage regulators, are designed to protect against minor power surges, spikes, and noise. They do not offer protection from the more serious power problems such as sags and brownouts, nor do they contain a battery backup for power. Line conditioners, voltage regulators and surge suppressors are often used with computer hardware and other sensitive equipment.

UPS SYSTEMS

To protect against data loss due to interruptions or fluctuations in commercial power, many critical use areas use an Uninterruptible Power Supply (UPS) on all computers and attached equipment. An UPS is a specially-designed power supply, with batteries that can supply power to the computer for short periods. If the commercial power fails, the computer runs off the UPS batteries until it can be shut down properly.

To enhance the protection offered by a UPS, many software versions include a built-in UPS monitoring feature. This feature provides an interface between the UPS unit and the operating system. Through this interface, the UPS can signal its status to the operating system and thereby coordinate an orderly shutdown.

The UPS provides more complete power protection but not all UPS systems are alike. Early UPS units were designed simply to provide backup power in the event of a blackout, leaving the equipment exposed to other types of power irregularities. Newer UPS systems are designed to protect against the full range of power problems. UPS units use an internal battery or set of batteries that produce AC power with an inverter. The inverter takes the DC battery voltage and changes it to AC.

UPS systems can be grouped into off-line, on-line, and hybrid. A newer type of power system is the intelligent power supply (IPS).

An off-line UPS is also known as a standby power system (SPS). It operates by switching from commercial power to battery power when the commercial power drops below a certain voltage level. The inverter is not powered, except during power outages.

Since the load receives power directly from the utility line as long as it is available, the load remains exposed to such power disturbances as spikes, noise, surges, and brownouts. Because of this, SPS-type systems should be used with a line conditioner or voltage regulator.

Another disadvantage of an SPS is that it takes a certain amount of time for the unit to take over after it senses that the voltage has dropped. Sensitive equipment may not be able to ride through this transition period without being affected. The time that an off-line UPS takes to make the switchover is critical. A switching time of four milliseconds or less is needed for electronic loads such as computer or control hardware. The major advantage of an off-line UPS is its low cost.

An on-line UPS acts as another source of power. It continuously converts commercial AC power to DC charging power to keep the battery charged. Its inverter uses the DC power from the battery to create new, clean AC power in the form of a steady 120-volt, 60-Hz sinewave. Critical loads run off the power generated by the UPS at all times.

The on-line UPS uses the commercial line voltage to keep its batteries charged, but never to supply power directly to the load. There is complete isolation of the load from the commercial power line and any power fluctuations. There is no switching of the load between commercial and battery power.

A hybrid UPS uses an off-line SPS with electronic or ferroresonant conditioning/filtering to smooth the transition from utility to inverter power. The quality of output power from a hybrid UPS depends on its filtering and conditioning capabilities. Many of these systems are advertised as line interactive, no break, load sharing and bidirectional. Sometimes hybrids are even labeled as on-line products although they do not function the same as on-line UPS units. Hybrid and off-line UPS units do not regenerate power continuously so the load receives raw or partially filtered utility power. Sensitive equipment can suffer from potential damage from this raw or partially filtered utility power.

Intelligent Power Systems (IPS) are specifically designed to protect computer systems. They go beyond UPS protection by adding software capabilities that work with the computer or network operating system. The interface to the operating system is more than a status indicator of power or battery charge. Built into the software are UPS monitoring functions to provide the network or workstation with an orderly, automatic, and unattended shutdown.

A true regenerative, on-line, sinewave UPS provides the most complete protection while maintaining a solid output during any power interruption. These type of systems are the most expensive and may not be able to handle all the loads.

Many UPS systems offer some protection against spikes, surges, and sags, but disregard brownouts. Off-line and other non-regenerative UPS units do not perform well during sustained low voltage conditions. Switching times for these UPS units tend to increase as the utility voltage decreases. A unit with a 5-millisecond transfer time at 120 VAC may exceed twenty milliseconds at 100 VAC. A brief period of low voltage precedes most blackouts and this places the load at greater risk.

Off-line and hybrid UPS units may also sense a brownout as a blackout and prematurely switch to battery. During a sustained brownout, an off-line UPS can discharge the battery and lose power even though the utility power is still on.

UPS units are available with either sinewave or squarewave outputs. These may be modified or approximated in some way. Sinewave is usually considered best since it is the same waveform provided by the utility companies and is the waveform that most equipment is designed for. Sinewave is better since some hardware may be affected by both linear or RMS currents and nonlinear or peak currents.

A squarewave output only approximates a sinusoidal waveform and puts stress on RMS-sensitive system elements, while starving peak-sensitive elements. This can cause excessive heating and hardware failures. The excess energy in squarewaves in the form of harmonics can affect electronic circuitry and cause data errors.

Computer grade sinewaves try to use enough squarewave increments to closely approximate a sinewave. This usually provides the needed RMS voltage and limits the peak voltages to equipment design values. But, these apparent sinewaves may not provide the precise timing needed by many computer monitors. The timing mechanism is referenced to the zero crossing of the utility sinewave. It regulates the scanning of the monitor. The imprecise screen scans from the zero crossing problem result in the screen appearing to swim or undulate.

The waveform from the UPS must also be a pure alternating current with no DC component. Even a small percentage of direct current in the output can saturate magnetic loads such as fans and transformers and make them inoperative.

The UPS must be able to start up all connected loads at a critical time. When the loads are energized again after a power anomaly, the initial inrush current is much higher than the normal operating power requirement. If additional equipment connected to the UPS is powered on when the demand for UPS power is already high, an overload condition may occur. The additional devices may not receive power, or other equipment attached to the UPS will be shut down. On-line UPS devices may require an inrush capacity as high as 1000%. On-line UPS units with inadequate inrush capacity cannot be used to power equipment up to the rated output of the UPS.

UPS units must have a power rating sufficient to support all of the loads that will require power from the unit. Power ratings are usually specified in terms of volt-amps (VA) or watts.

Extended operation on the UPS batteries in the absence of commercial power will shorten the life of the batteries. Most UPS systems have a maximum time before the unit switches back to commercial power.

The increasing need for reliable power quality in today's industrial and commercial facilities has resulted in the availability of a wide variety of power quality migration equipment, among them static and rotary UPS systems. The level of power protection provided by these systems depends on the maintenance of electrochemical storage batteries.

CHEMICAL BATTERIES

Lead-acid battery based ride-through systems are common. The first electric storage goes back to the Italian count Alessandro Volta, whose voltaic pile in 1800 was the world's first battery. Volta used a cell with round plates of copper and zinc as electrodes. Cardboard soaked in salt water was the electrolyte. Current flowed but this primary cell could not store power for any length of time.

When lead-acid batteries are regularly deep cycled to 80% or more of full discharge, their life is usually limited to less than 600 charges/discharges. When only shallow discharges are used, leaving 2/3 or more of the battery's full capacity, the number of charge/discharge cycles can go above 10,000. For most UPS ride-through applications, the discharges are relatively infrequent (less than 100 per year) and a short duration does not require complete discharge.

Most chemical batteries must be replaced every 6 to 7 years. The replacement and disposal cost must be factored into the total energy cost. Battery systems are heavy relative to the amount of power they can deliver and requiring significant power storage can weigh several tons.

Chemical-battery-based ride-through systems typically use sealed, maintenance-free automotive batteries with thicker lead plates and higher electrolyte levels. Smaller systems can provide 250-kW for 10 seconds while larger 10-MW, 10-second are available. A 1-MW system will use 240 lead-acid cells weighing 46,000 pounds. These are capable of delivering 1,800 amps for five minutes. Full output is achieved within two milliseconds of a utility power interruption.

Performance is limited by the lead-acid battery packs which are generally the most affordable option. More exotic batteries like nickel metal hydrid (NiMH) packs have also appeared.

The common 12-volt lead-acid battery has six cells, each containing positive and negative lead plates in an electrolyte solution of sulfuric acid and water. This proven technology is not expensive to manufacture and it's relatively long-lasting. But, the energy density of lead-acid batteries, the amount of power they can deliver on a charge, is poor when compared to NiMH and other newer technologies.

The United States Advanced Battery Consortium (USABC) is a Department of Energy program launched in 1991. Since 1992, USABC has invested more than $90 million in nickel metal hydrid batteries.

These batteries are much cheaper to make than earlier nickel battery types, and have an energy density almost double that of lead-acid. NiMH batteries can accept three times as many charge cycles as lead-acid, and work better in cold weather. NiMH batteries have proven effective in laptop computers, cellular phones, and video cameras.

NIMH batteries can take an electric vehicle over 100 miles, and are still several times more expensive than lead-acid. NiMH Batteries from Energy Conversion Devices were installed in GM's EV_I and S-10 electric pickup truck, doubling the range of each. Chrysler has also announced that NiMH batteries, made by SAFT of France would go in its Electric Powered Interurban Commuter (EPIC) vans, adding 30 miles to their range.

Other battery technologies include sodium-sulfur which was used in early Ford EVs, and zinc-air. Zinc appeared in GM's failed Electrovette EV in the late 1970s. Zinc-air batteries have been promoted by a number of companies, including Israel's Electric Fuel, Ltd. Zinc is inexpensive and these batteries have six times the energy density of lead-acid. A car with zinc-air batteries could deliver a 400-mile range, but the German postal service demonstrated that these batteries cannot be conventionally recharged.

Other battery types are more promising, including lithium-ion, which is used in a variety of consumer products. Lithium batteries could offer high energy density, long cycle life, and the ability to work in different temperatures. However, like the sodium-sulfur batteries in the Ford Ecostar, lithium-ion presents a fire hazard since lithium itself is reactive.

Plastic lithium batteries could prove to be very versatile. Bellcore is working on a lithium battery that would be thin and bendable like a credit card for laptop computers and cell phones. Each cell is only a millimeter thick. The plastic batteries are lightweight and have seen some preliminary testing in automotive applications.

Canadian utility Hydro-Quebec has been working with 3M on a lithium-polymer unit which could be the first dry electric vehicle battery. Like the Bellcore product, this dry battery uses a sheet of polymer plastic in place of a liquid electrolyte. Also working on this technology is a team at Johns Hopkins University. This is also a plastic battery that can be formed into thin, bendable sheets. These batteries also contain no dangerous heavy metals and are easily recycled.

KINETIC ENERGY STORAGE

Batteryless or kinetic energy storage technology has been in use for many years, primarily for load isolation and conditioning. Flywheel systems must have stability and low frictional losses. Magnetic bearings are one solution since they allow a flywheel's rotating shaft to float, with no surface-to-surface contact. Previous magnetic bearing research focused on high-power active magnetic bearings that use too much energy to be practical. Hybrid magnetic bearings use permanent magnets and provide correction and stability with much smaller and more energy efficient electronic components. Alternatives to magnetic bearings include ceramic bearings that provide nearly frictionless rotation, despite the surface-to-surface contact.

Advances in component technologies have benefited flywheel systems. Over the last 30 years, the tensile strength of graphite composite materials has increased some five-fold, while the cost per pound has dropped over 90%. The development of Kevlar marked a materials breakthrough in flywheel system development. More recently, T1000 graphite composites with tensile strengths up to a million pounds have allowed flywheels to approach 100,000 rpm, dramatically increasing their energy storage capacity.

Magnetic materials have also seen a marked improvement in recent years with the development of rare earth magnets such as samarium cobalt-17 in the mid-1970s and neodynium-iron-cobalt (NdFeB) in 1983. The latter is used extensively by the automotive industry in alternators. Materials costs for NdFeB are quite low, since neodynium is the third most common rare earth element, and cobalt is used only in small amounts.

A flywheel-coupled motor-generator is used in International Computer Power's Dynamic Energy Storage System (DESS). This is a kinetic storage system that provides system ride-through. These units can be used to replace UPS battery backup systems in standby engine generator applications. Kinetic-storages can also be used with an existing battery storage system to prevent unnecessary battery system discharging during short-duration power anomalies.

Over 99% of all power disturbance are a few cycles to a few seconds long. Many times an UPS will be used for the few seconds that it takes for a standby engine generator to start and become operational. Storage batteries are typically used to provide the necessary backup power to the load during this short time period.

These battery systems need constant monitoring, maintenance and service. Batteries also require a large environmentally controlled storage area and special disposal procedures, which increases the operational costs through regulatory compliance. Each discharge and charge cycle also decreases the useful cell life.

Some power protection systems combine slow-speed conventional flywheels with other energy storage and generation technologies. These include synchronous generators, diesel or natural gas engines with flywheels that rotate at less than 6,000 rpm.

The Statordyne system uses hydraulic storage with back-up generation. The system uses a synchronous motor/generator that provides power factor correction by running as a synchronous condenser while the utility is supplying power. It switches to a generator when the utility power is interrupted.

Energy is stored in a mechanical flywheel that keeps the generator's speed at about 1,777 rpm for the first 100 to 200 milliseconds until a hydraulic motor is engaged to keep the generator's shaft spinning at 1,800 rpm until a diesel or natural gas generator starts and takes over. These systems range from 100- to 800-kW, and cost about $1,000/kW depending on the size and features.

The Holec system also uses flywheel-inertia storage with back-up generation. This system integrates low-speed flywheel technology with a synchronous generator to protect facilities from power outages 2 to 3 seconds in duration, with a diesel or natural gas engine taking over for protection from longer power outages. The Holec system is installed in about 300 facilities worldwide. The system uses a 3,600-rpm induction motor, with a rotating stator that is mechanically attached to the rotor of an 1,800-rpm motor/generator. Under normal conditions, both motors operate at full speed and the rotor of the induction motor turns at 3,600 + 1,800 rpm for an overall rotating speed of 5,400 rpm. It acts like an energy storage flywheel.

When an electric power interruption occurs, the induction motor is reconfigured to operate as an eddy current clutch. It transfers energy from itself to keep the generator's shaft at full speed. These systems are available in sizes ranging from 100 to 2,200 kVA and cost about $700 per kVA for a 1-MW system.

SUPERCONDUCTING MAGNETIC ENERGY STORAGE

This technology stores electrical power in a coil of superconducting wire submerged in liquid helium. It may be used for both ride-through and energy storage systems.

Since the late 1960s, over 20 development projects have begun, with some outstanding results. Storing large amounts of energy this way is feasible and the ability to deliver high bursts of power make it attractive for facility ride-through systems.

This technology stores electrical energy without intermediate conversions to mechanical or chemical energy. A superconducting magnetic coil is immersed in liquid helium at 4.2°K (-455.5°F), causing its resistance to DC current to fall to zero. High electrical currents can be sent to the coil and the current will circulate without any losses, until it is diverted from the coil to the facility electrical system. The very cold temperatures used require continuous cooling. The energy used by the cooling systems is about 25-kW for a moderately sized unit that stores about 0.28-kWh.

Superconductivity, Inc., has been shipping commercial systems since 1992 with over a dozen ride-through systems currently installed. Systems are available for about 1,000 per kW. The company is now part of American Superconductor, Inc. located in Westborough, MA.

POWER QUALITY PERFORMANCE EVALUATION

Power quality performance evaluation involves finding out how good the power quality is and what performance improvements should be made. In the past, power quality was not recognized as a major problem since most of the load was rotating motors and resistive heating or lighting which are much more tolerant and insensitive than the present load mix. Monitoring and instrumentation was much less developed and incapable of capturing most transient power events. The knowledge of utility and site generated events was also limited.

As far as the utility was concerned, they were the supplier of utility grade raw power. Utilities relinquished responsibility at the Point of Common Coupling (PCC). The PCC is the point at which the ownership of the electrical wiring changes from the utility to the facility.

The customer was left with the responsibility to condition their equipment, if required, to accommodate normal power anomalies. Most customers were subject to these events in varying degrees, but this situation had always existed and over time there was some performance improvement in the frequency of interruptions.

This performance criteria lacked the impact of sags, surges, transients and other electrical events. The actual criteria was strictly based on the percent of power availability for a facility.

POWER MONITORING AND CONTROL

The driving factors in early industrial power monitoring, exclusive of control, were and still are, reduction in labor and timeliness of recording meter readings. In large physical plants, meter output needed to be transformed into a signal for transmission, redisplay and recording at a centralized location (Figure 4-1).

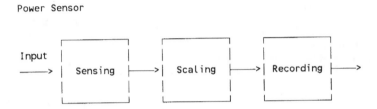

Figure 4-1. Power metering

By the late 1970s, minicomputers were being used in relatively elaborate power monitoring and control systems. These systems used analog metering and were installed in energy intensive industrial processes. These first power monitoring and control applications were custom proprietary systems with high development costs for software. The market for such systems fell as the price of oil dropped.

By the late 1980s, remote power monitoring systems appeared with PCs running extensions to the DOS operating system. This allowed simple multitasking capability so a PC could scan a serial line running a multi-drop protocol to field instruments. PLCs were used and there was some level of operator display with data logging and query support. These systems were an order of magnitude less in cost than the systems of a decade earlier. The market was driven by large industrial facilities whose physical scale and size made remote meter reading a necessity. Another factor was the need for industrial monitoring systems with some SCADA functions for recently empowered cogenerators who started to sell power back to the utilities, who were now required by the Public Utilities Regulatory Policies Act of 1978 (PURPA) to purchase power from them.

In 1996 a new generation of industrial power monitoring and control systems began to appear (See Table 4-2). These systems were driven by the availability of powerful low-cost PCs and embedded microprocessors, a large installed base of computer networking technologies such as Ethernet and industrial acceptance of standard operating systems like NT that operate on low cost personal computers.

Table 4-2. PM&C Systems

Company	On-line Help	Trending	Logging	Annunicated	Report Generator
Power Measurement Limited	X	X	X	X	X
E-Mon			X		
Scientific Columbus			X		X
Operation Technology	X	X	X	X	X
Square D	X	X	X	X	X
Cutler Hammer	X	X	X	X	X
GE/Multilin	X	X	X	X	X
Electro/Indus Gaugetech	X	X	X	X	
Allen Bradley	X	X	X	X	
Jomitek		X	X	X	
ABB	X		X	X	
GE Company	X	X	X		
Siemens	X	X	X	X	

Companies like Power Measurements began offering systems that could be deployed at a single substation or over a LAN or independent network. These systems could be used to link a large enterprise over a wide area network to support corporate engineering and procurement functions.

Other companies like Allen-Bradley, who until this time had sold rebranded Power Measurements OEM devices, developed new intelligent metering and local control systems. Along with RS Software, a part of Allen-Bradley's parent company Rockwell International, Allen-Bradley began to link power monitoring with automation.

These second generation systems were divided into two basic types. An entry level system was primarily a monitoring system for remote metering. There were also modular systems for power monitoring as well as supervisory control, with direct distributed control in the individual metering loops (Figure 4-2).

The newer systems provide information for energy procurement decision making and integrate power acquisition from multiple suppliers. They provide distribution system coordination, power capacity planning and allow integrated power and production control planning.

As power quality and reliability issues continue to grow in importance, these power monitoring and control systems become critical in providing the needed information and automatic control to support business and financial objectives in the acquisition and distribution of electrical energy.

PM&C products include utility and industrial systems. The power monitoring and supervisory control systems used by utilities are usually part of the utility's energy management systems. Because of the physically distributed nature of the monitoring and control requirements these systems are characterized by the term SCADA (Supervisory Control and Data Acquisition). These systems are usually interfaced with IMS systems (Information Management Systems) and make up the overall EMS (Energy Management System).

Most industrial PM&C systems are stand-alone systems and are not incorporated into process control, shop floor control or other factory information systems, including production scheduling, MRP (Material Requirements Planning) or ERP (Enterprise Requirement Planning).

Entry level monitoring systems evolved from traditional power

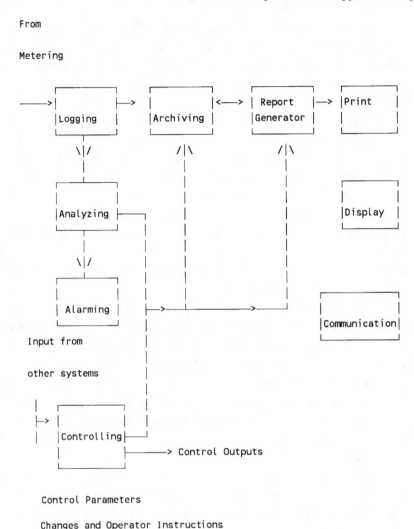

From

Metering

Figure 4-2. Modular power monitoring and control

monitoring systems made up of mechanical meters and recorders. The newer monitoring systems add automation to their basic functionality and are primarily electronic. The main function of these monitoring systems is to provide a central readout point for a number of local metering devices. Installation of these systems provides a means of obtaining rapid status and update information to correct problems as well as current data on energy consumption for the purposes of cost allocation and chargeback.

Higher levels of functionality include supervisory control either in a distributed or centralized mode of operation. The hardware configuration may be similar but the software needed to implement supervisory control is significantly more complex.

Systems may be embedded and non-embedded. Embedded systems are intelligent monitoring and control devices with software that has been burned into ROM (firmware). Programming is limited and is done with a keypad or programming unit.

Non-embedded systems use a stand-alone or networked computer to load software, configure and control more advanced functions. Non-embedded systems often use a master-slave scheme, where remote units report or summarize data from other units without the need for centralized supervisory control.

Other systems are based on distributed monitoring and control where all units, meters and monitoring points can operate independently and communicate interactively. They can also report hierarchically. There may also be software for dynamic purchasing, energy allocation, and interaction with utility computers for such functions as automatic dynamic wheeling.

POWER QUALITY EVALUATION

One approach to power quality evaluation is based on the analysis of circuits and equipment operating specifications. The environment must be well understood and documented. Equipment manufacturers often exaggerate equipment abilities and most products are purchased on specifications, not on test results.

Another approach to power quality evaluation is based on the facility sensitivities to power anomalies. These sensitivities are monitored and include outages, sags, surges, imbalance, transients, harmonics and other power problems in terms of magnitude, duration, and other applicable parameters. Events must be continuously monitored so that they can be grouped as acceptable or non-acceptable to a defined set of rules. These rules could be graphical such as those developed by the Computer Business Equipment Manufacturers' Association (CBEMA) or the IEEE P-1159 committee.

Under this approach, the utility is contracted to supply power that satisfies the defined criteria. Monitoring is used to verify compliance.

The performance based approach is based on a cause and effect scheme. The key is to relate each process or equipment malfunction to a power event. This can be accomplished by a log of events that can be correlated with equipment malfunctions. Log entries include the cost of each event and the frequency of each event. A cost/benefit analysis can be performed and changes can be made to improve performance.

References

Arthur, Bob, "Testing Reveals Surprising K-factor Diversity," *Electrical Construction and Maintenance*, Vol. 92 No. 4, April 1993, pp. 51-54.

Bishop, Martin T., and John F. White and Robert B. Morgan, "Steel Mill Tackles Power Quality and Power Factor Problems," *Electrical Construction and Maintenance*, Sept. 1996, Vol. 95 No. 9, pp. 48-56.

Cogdell, J.R., *Foundations of Electrical Engineering*, Prentice Hall: Englewood Cliffs, NJ, 1990.

Frank, Jerry, "The How and Why of K-factor Transformers," *Electrical Construction and Maintenance*, Vol. 92 No. 5, May 1993, pp. 79-82.

Herrick, Clyde N. and Kieron Connolly, *Electrical Control Systems for Heating and Air Conditioning*, Lilburn, GA: The Fairmont Press, Inc. 1998.

Liebing, Edward and Ken Neff, *Netware Server*, New York: McGraw-Hill Inc., 1990.

Mather, Randy, "The Keys to Evaluating Power Line Loads," *Electrical Construction and Maintenance*, Vol. 99 No. 3, July 1999, pp. 6-7.

Moravek, James M. and Edward Lethard, "The K-factor: Clearing Up its Mystery," *Electrical Construction and Maintenance*, Vol. 92 No. 6, June 1993, pp. 65-67.

Motavalli, Jim, *Forward Drive*, San Francisco, CA: Sierra Club Books, 2000.

Newcombe, Charles and Jeanine Katzel, "Understanding the Two Readings of Power Factor," *Plant Engineering*, Vol. 50 No. 2, Feb. 1996, pp. 120-123.

Ricketts, Jana, Editor, *Energy and Environmental Visions for the Millennium*, Lilburn, GA: The Fairmont Press, Inc., 1998.

Turner, Wayne C., *Energy Management Handbook*, The Fairmont Press, Inc.: Lilburn, GA, 1997.

CHAPTER 5

FUEL CELL POWER

FUEL CELL TECHNOLOGY

There has been steady and substantial improvements in fuel cell technology. Today, fuel cells are very newsworthy. Fuel cells were first demonstrated in 1839 but have not been widely applied because the technology was expensive and other technologies have dominated the power marketplace. However, their exceptional environmental performance and their notable efficiencies have recently caused renewed interest, buoyed by advances in engineering and material science. Fuel cells may provide a real alternative for producing power.

Fuel cells convert the chemical-energy of fuels directly into electricity. The principle of the fuel cell was demonstrated by William Grove in 1839. Sir William Robert Grove was trained as a barrister and became a judge. He got involved in patent cases and would suggest improvements. He made several important improvements to the design of storage batteries and showed how the combination of hydrogen and oxygen could be used to produce electricity.

His work was based on the idea that it should be possible to reverse the electrolysis process and produce electricity rather than use electricity to cause the chemical changes needed for metal plating. It could find no practical application during the inventor's lifetime and the actual uses for this gas battery did not interest him.

Fuel cells can be compared to a car battery, in that hydrogen and oxygen are combined to produce electricity. Batteries store their fuel and their oxidizer internally, meaning they have to be periodically recharged. A fuel cell can run continuously as long as it has fuel and oxygen. The hydrogen can be extracted from a fuel like methanol or gasoline. It is fed to the anode, one of two electrodes in each cell.

The process strips the hydrogen atoms of their electrons, turning them into positively charged hydrogen ions, which then pass through an electrolyte (See Figure 5-1). Depending on the type of fuel cell, this may be phosphoric acid, molten carbonate, or another substance. The ions are sent to the second electrode, known as the cathode.

The negatively charged electrons, which cannot travel through the electrolyte, move to the cathode through a conductive path or wire. This movement produces an electric current. The amount of current is determined by the size of the electrodes. At the cathode, the electrons combine with oxygen to produce the fuel cell's major by-product, water. The other by-product is waste heat, which in some applications can be captured and reused in a processed cogeneration.

A single cell generates only one volt, but the cells are grouped in stacks to produce higher voltages. In a pure hydrogen fuel cell, emissions are nil, but some release of pollutants occurs in cells that reform or extract, the hydrogen from a fossil fuel. Unlike an internal combustion engine with its noise, heat, and moving parts, the fuel cell is an enclosed box, with no moving parts and no noise.

By 1900 scientists and engineers were predicting that fuel cells would be common for producing electricity and motive power within a few years. That was more than 100 years ago. Today, gas turbine combined cycle powerplants are generally the most efficient type of engine available for power generation.

Fuel cells are in about the fifth cycle of attempts to turn them into commercial products. In the past these attempts failed, but a few companies continued development. It was General Electric that did most of the pioneering work on proton exchange membrane (PEM) fuel cells in the early 1960s. GE developed its PEM cells for the Gemini space program. These cells cost hundreds of thousands of dollars for each kilowatt generated, so GE saw no practical applications and let many of its patents run out. GE recently announced it was get-

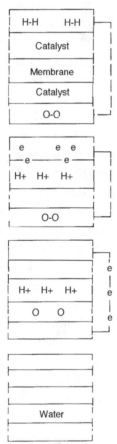

Figure 5-1. Fuel cell operation

ting back into fuel cells as a distributor for other companies' products.

This time around we have more advanced materials than were available in the 1960s when the last wave of development occurred. It appears that much has changed. One trend is to do things directly. The less expensive solution does away with any intermediate steps. The focus is on efficient and clean technologies that use readily available fuels.

FUEL CELL CHARACTERISTICS

A fuel cell is an energy conversion device that generates electricity and heat by electrochemically combining a gaseous fuel and an oxidant gas across an ion conducting electrolyte. The chief characteristic of a fuel cell is its ability to convert chemical-energy directly into electrical energy, giving high conversion efficiencies.

Fuel cells have much lower carbon dioxide emissions than fossil fuel based technologies for the same power output. They also produce negligible amounts of SO_x and NO_x, the main constituents of acid rain and photochemical smog.

Several types of fuel cells are being developed around the world, the chief difference between each being the material used for the electrolyte and the operating temperature. The types of fuel cells include solid oxide, molten carbonate, phosphoric acid, polymer, direct alcohol and alkaline (See Table 5-1).

All fuel cells use an electrolyte between the two electrodes. The different types of electrolytes have very different properties and the different fuel cell types have been built around them and are mostly named after the electrolyte.

When fuel cells transform the energy stored in a fuel into electricity and heat, the fuel is not burned in a flame but oxidized electrochemically. This means that fuel cells are not constrained by the law that governs heat engines, the Carnot limit, which specifies the maximum theoretical efficiency that a heat engine can reach. Their efficiency increases with a partial load.

A fuel cell works similar to a battery. In a battery there are two electrodes which are separated by an electrolyte. At least one of the electrodes is generally made of a solid metal. This metal is converted to

Table 5-1. Types of Fuel Cells

Type	Abbre-viation	Operating Temperature	Characteristics Uses
Alkaline	AFC	50-250°C	Efficient, costly; Used in space vehicles
Direct Alcohol	DAFC	50-100°C	Buses, cars, appliances, small CHP, work needed in membrane development
Molten Carbonate	MCFC	600°C	Large CHP, range of fuels
Phosphoric Acid	PAFC	200°C	Medium CHP, steam generation
Polymer Electrolyte	PEFC	50-100°C	Buses, cars
Solid Oxide	SOFC	500-1000°C	All sizes of CHP

another chemical compound during the production of electricity in the battery. The energy that the battery can produce in one cycle is limited by the amount of this solid metal that can be converted. Larger batteries have more metal exposed to the electrolyte.

In the fuel cell the solid metal is replaced by an electrode that is not consumed and a fuel that is continuously replenished. This fuel reacts with an oxidant such as oxygen from the other electrode. A fuel cell can produce electricity as long as more fuel and oxidant is pumped through it.

There are now several types of fuel cells that appear to be most promising. None of these fuel cells are in commercial production. Solid oxide fuel cells (SOFCs) are most likely to be used for large and small electric powerplants above 1-kW. The direct alcohol fuel cell (DAFC) is more likely to be a battery replacement for portable applications such as cellular phones and laptop computers.

The polymer electrolyte fuel cell (PEFC) may be the most practical in a developed hydrogen economy. The DAFC may be much simpler than the PEFC making it better for vehicular applications. The much higher efficiency of the SOFC and it's ability to use most any fuel make it a contenter for vehicular applications. The start-up time of the SOFC may be overcome by using supercapacitor batteries for the first few minutes of operation.

Several fuel cells are in limited production. The polymer electrolyte fuel cell (PEFC) is in limited commercial production. The phosphoric acid fuel cell (PAFC) is being produced for medium sized electric powerplants. The alkaline fuel cell (AFC) has been produced in limited volumes since the early space flights. The SOFC is considered to be superior to the PAFC and would likely replace it in time. The molten carbonate fuel cell was believed to be best for electric powerplants due to the potential problems of the SOFC. Since it appears these problems may be solved, development of the MCFC may cease.

The alkaline fuel cell (AFC) has been used since the early space applications where hydrogen and oxygen are available. Using carbon dioxide scrubbers allows these fuel cells to be operated on hydrogen and air.

FUEL CELLS FOR ELECTRIC POWER PRODUCTION

The trend is to deregulate the production of electric power. Deregulation should promote combined heat and power (CHP), also known as cogeneration. CHP conserves fuel by using the thermal-energy that is produced from generating electricity. Since thermal-energy cannot be piped over long distances, CHP powerplants generally tend to be much smaller than the present units.

Fuel cells may be favored for these smaller electric powerplants. Fuel cells are ideal for electric power production since electricity is both the initial and final form of energy that is produced. They need to produce reasonable efficiencies in 30-kW sizes, run quietly, require infrequent maintenance and emit little pollution. Electricity is used by many high technology portable devices. The current batteries used in many devices do not have a very long life. Fuel cells could provide continuous power for these devices. Every week or month a new supply of liquid fuel could be injected into the fuel cell.

Alkaline Fuel Cells

Alkaline fuel cells (AFCs) use hydrogen and oxygen as fuel. The alkaline fuel cell is one of the oldest and most simple type of fuel cell. This is the type of fuel cell that has been used in space missions for some time. Hydrogen and oxygen are normally used as the fuel and oxidant. The electrodes are made of porous carbon plates which are laced with a catalyst to accelerate the chemical reactions. The electrolyte is potassium hydroxide. At the anode, the hydrogen gas combines with hydroxide ions to produce water vapor. This reaction results in electrons that are left over. These electrons are forced out of the anode and produce the electric current. At the cathode, oxygen and water plus returning electrons from the circuit form hydroxide ions which are again recycled back to the anode. The basic core of the fuel cell consisting of the manifolds, anode, cathode and electrolyte is called the stack.

Alkaline fuel cells use a solution of potassium hydroxide in water as their electrolyte, making them sensitive not to CO as the SPFC is, but to CO_2. The alkaline fuel cell cannot operate with carbon dioxide in either the fuel or oxidant. Even a small amount of carbon dioxide in the air is harmful. Carbon dioxide scrubbers have been used to allow these fuel cells to operate on air. The cost of the scrubber is considered to be reasonable.

Oxygen has traditionally been used as the oxidant in the system. This has led to few uses outside aerospace, although some of the first experimental vehicles were powered by AFCs. They use comparatively cheap materials in their electrodes but are not as power dense as SPFCs, making them bulky in some situations.

The alkaline fuel cell has been used with great success in the past in space missions, dating back to the Apollo and Gemini missions in the 1960s. It is still in use in the Space Shuttle today and provides not only the power but also the drinking water for the astronauts. The space-qualified hardware made by United Technologies Corporation is very expensive, and until recently it was not thought that it would be useful in other applications. However, in July 1998 the Zero Emission Vehicle Company (ZEVCO) launched its first prototype London taxi based on the technology. Using a 5-kW fuel cell and about 70-kW of batteries, the hybrid taxi is a zero emissions vehicle capable of operating in cities.

This type of fuel cell operates at various temperatures, 250°C was used in space vehicles. The DC efficiency is as high as 60% at rated power and since there are low system losses, the partial load efficiency can be even higher.

Direct Alcohol Fuel Cells

Several companies are working on direct alcohol fuel cells (DAFCs). JPL has been working on DMFC since 1992. Many of the increases in efficiency and power density are as a result of their efforts. The operating temperature is 50-100°C, which is ideal for small to mid-size applications. The electrolyte is a polymer or a liquid alkaline.

In 1999 there was a movement away from developing the PEFC in favor of the DAFC. In the DAFC, methyl (DMFC) or ethyl (DEFC) alcohol is not reformed into hydrogen gas but is used directly in a simple type of fuel cell. This type of fuel cell was bypassed in the early 1990s because its efficiency was below 25%. Most companies pursued the PEFC because of its higher efficiency and power density.

Efficiencies of the DMFC are now much higher and projected efficiencies may reach 40% for DC automobile applications. Power densities are over 20 times as high now as in the early 1990s. The DMFC could be more efficient than the PEFC for automobiles that use methanol as fuel.

Fuel passing from the anode to the cathode without producing electricity is one problem that has restricted this technology. Energy Ventures claimed in 1999 that it has solved this cross-over problem. Another problem is the chemical compounds that are formed during operation that poison the catalyst.

The Direct Alcohol Fuel Cell appears to be most promising as a battery replacement for portable applications such as cellular phones and laptop computers. There are working DMFC prototypes used by the military for powering electronic equipment in the field. Small units for use as battery replacements do away with the air blower and the separate methanol water tank and pump. These fuel cells are not very different than batteries in construction.

One concern is the poisonous aspect of methanol-methyl alcohol.

Methanol can be replaced by ethanol. Several companies are working on DEFC. The power density is only 50% of the DMFC but this may be improved.

Direct Methanol Fuel Cells (DMFC)

This type of fuel cell is based on solid polymer technology but uses methanol directly as a fuel. It would be a big step forward in the automotive area where the storage or generation of hydrogen is one of the big obstacles for the introduction of fuel cells. Prototypes exist, but the development is at an early stage. There are principal problems, including the lower electrochemical activity of the methanol as compared to hydrogen, giving rise to lower cell voltages and, hence, efficiencies. Also, methanol is miscible in water, so some of it is liable to cross the water-saturated membrane and cause corrosion and exhaust gas problems on the cathode side. The Direct Methanol Fuel Cell is being worked on at Siemens in Germany, the University of Newcastle and Argonne National Laboratory.

There are also efforts to develop a low-temperature SOFC (500°C) that would also allow the use of methanol directly, as well as using stainless steel components. The Imperial College in London is active in this area.

Molten Carbonate Fuel Cells

Molten Carbonate Fuel Cells (MCFCs) use an electrolyte that is a molten alkali carbonate mixture, retained in a matrix. They operate at a temperature of about 650°C and useful heat is produced.

The cathode must be supplied with carbon dioxide, which reacts with the oxygen and electrons to form carbonate ions, which convey the ionic current through the electrolyte. At the anode these ions are used in the oxidation of hydrogen. This also forms water vapor and carbon dioxide to be conveyed back to the cathode. There are two ways to accomplish this: by burning the anode exhaust with excess air and removing the water vapor before mixing it with the cathode inlet gas or by separating the CO_2 from the exhaust gas using a product exchange device.

The fuel consumed in an MCFC is usually natural gas, although this must be reformed in some way to create a hydrogen-rich gas to feed the stack. An MCFC produces heat and water vapor at the anode, which can be used for the steam reformation of methane. This means it is fundamentally more efficient than a cell requiring external fuel processing.

The MCFC may be used for large scale power generation. One reason is the necessity for auxiliary equipment, which can make smaller operations uneconomical. Fuel Cell Energy is working on MCFC 300-kW, 1.5-MW and 3-MW power generation units. This technology cannot be scaled down below 300-kW because of their need for significant amounts of auxiliary equipment such as pumps.

There is no requirement for catalysts as needed in low temperature fuel cells and the heat generated can be used for internal reformation of methane, a bottoming cycle and for fuel processing and cogeneration. This increases the overall efficiency of the generating system.

The Molten Carbonate Fuel Cell has been under development for almost 20 years as an electric powerplant. The operating temperature of 600-650°C is lower than the SOFC. It is significantly more efficient that the PAFC and has the advantage of reforming inside the stack. One disadvantage is the corrosiveness of the molten carbonate electrolyte. Large powerplants using gas turbine bottoming cycles to extract the waste heat from the stack could be up to 60% efficient when operating on natural gas. If problems with the SOFC are solved, work on the MCFC may fade.

It is unclear whether hydrogen fuel will be widely used. This is because solid oxide fuel cells may become popular and these can cleanly convert renewable hydrocarbon fuels. The Solid Oxide Fuel Cell may be the most promising technology for small electric powerplants over 1-kW.

Phosphoric Acid Fuel Cells

The Phosphoric Acid Fuel Cell (PAFC) is one of the oldest and most established of the fuel cell technologies. It has been used in several power generation projects. It has a phosphoric acid electrolyte and can reform methane to a hydrogen-rich gas for use as a fuel with the waste heat from the fuel cell stack. This heat may also be used for space heating or hot water.

It operates at temperatures around 200°C, so the waste heat is not hot enough to be used for cogeneration. It is possible to use alcohols such as methanol and ethanol as fuels, though care must be taken to avoid poisoning the anode by carbon monoxide and hydrogen sulfide which may be present in the reformed fuels. This results in a gradual reduction in performance and the eventual failure of the cell.

Japan has done some advanced PAFC research and design and has powerplants from a few kilowatts to a few megawatts in operation. Toshiba, Fuji and Mitsubishi and others are pursuing this technology. Japan's lack of natural resources is forcing this technology on the market at a higher price than would be possible in other countries.

The Phosphoric Acid Fuel Cell has been under development for almost two decades as an electric powerplant. While it has a lower real efficiency than the MCFC or SOFC, its lower operating temperature of 160-220°C is considered almost ideal for small and midsize powerplants. Midsize 200-kW AC powerplants are 40% efficient and large 10 MW units are 45% efficient when running on natural gas. These efficiencies are similar to the PEFC.

Polymer Electrolyte Fuel Cell

The Polymer Electrolyte Fuel Cell (PEFC) is considered the fuel cell of the hydrogen economy. Automobiles would emit pure water from their tailpipes.

PEFC systems would extract hydrogen from hydrocarbon fuels such as methanol or natural gas. The efficiency of the PEFC when running on hydrogen and no air pressurization is high but practical systems that use fuel reforming and air compression suffer in efficiency. Small 30-kW AC powerplants are likely to be 35% fuel to electricity efficient with 200-kW units at 40% and large units at 45%. An automobile powerplant with an electric motor would have an efficiency of about 35%.

There has been some progress in storing hydrogen in different materials such as hydrides or carbon. This would eliminate the need for a reformer but unless there is a hydrogen pipeline system, the hydrogen would have to be produced locally at service facilities. This is more likely to be done at larger metropolitan facilities.

The PEFC operates at 80°C which makes it useful for small applications and allows less expensive materials to be used. A catalyst is required to promote the chemical reaction at these low temperatures. The platinum catalysts used in the stack makes this type of fuel cell expensive, but new techniques for coating very thin layers of catalyst on the polymer electrolyte have reduced the cost of the catalyst to about $150 per unit.

PEFC can only use hydrogen for fuel and hydrocarbon fuels must be reformed carefully since small amounts of carbon monoxide can damage the catalyst. If a reformer is used, this requires a few minutes of warm up time. Stored hydrogen is used in the start-up phase. A liquid cooling system is required.

PEFCs larger than 1-kW are usually pressurized to increase the chemical reaction at the low temperatures involved. Air compressed to about 3 atmospheres or higher is used to increase the power density of the fuel cell. On small systems this results in a significant loss of efficiency. The air compressors also add more complexity to the fuel cell. On automobiles and buses two air compressors are often used, a turbocharger and a supercharger. The DAFC could replace the PEFC once these problems are solved.

Solid Oxide Fuel Cells

Advances in modern ceramic technology and solid-state devices are pushing the development of a range of efficient units. Many ceramics can be tailored to display electrical properties unattainable in their metallic or polymeric counterparts. These materials are called electroceramics. One group of electroceramics, the fast oxygen ion conductors, are used in devices such as oxygen sensors, oxygen pumps, exhaust catalysts and Solid Oxide Fuel Cells (SOFCs).

A SOFC uses yttria-stabilized zirconia as its electrolyte, between the anode and the cathode. It runs at a temperature of about 1,000°C. The heat produced can be used in cogeneration applications or in a steam turbine to provide more electricity than that generated from the chemical reaction within the fuel cell. This is known as a bottoming cycle. Several different fuels can be used, including pure hydrogen, methane and carbon monoxide. The nature of the emissions from the fuel cell will vary correspondingly with the fuel mix. Many SOFCs use a separate pre-reformer as opposed to an integral reformer. SOFCs generally use carbon monoxide as a fuel. There are three basic designs of SOFC: tubular, planar and monolithic.

Westinghouse has been working on a tubular form of SOFC. The tubular units operate with the fuel on the outside surfaces of a bundle of tubes. The oxidant is on the inside and the tube itself is composed of the electrolyte and electrode sandwich.

The tubes have a high electrical resistance but are simple to seal. Other companies such as Global Thermoelectric are working on planar SOFCs made of thin ceramic sheets which operate at less than 800°C. The thin sheets have a low electrical resistance and offer high efficiencies. Less expensive materials can be used at the lower temperatures and help the SOFC to reach commercial markets.

Planar SOFCs are being developed by several companies, including Siemans and Fuji Electric. In these units the cells are flat plates bonded together and form a stack. The advantage of this over the tubular system is its relative ease of manufacture. The lower ohmic resistance of the electrolyte results in reduced energy losses.

Siemens Westinghouse in Germany is working on tubular SOFCs operating at 1000°C. In 1998, Siemens halted work on its own planar solid oxide fuel cells and bought out Westinghouse's gas turbine and tubular solid oxide fuel cell division. Siemens planar design suffered from leaky seals of it's window frame type design which had 16 small SOFC cells in each layer. Siemens is also working on DMFC for automobiles and PEFC for specialty applications. Sulzer in Germany is working on a 3-kW SOFC for CHP.

Monolithic SOFCs are also in development that use a honeycomb structure. Tests indicate that this form of fuel cell may be one of the most efficient. They are capable of efficiencies between 50 to 60%. High-grade waste heat is produced, for combined heat and power (CHP) applications and internal reforming of hydrocarbon fuels is possible.

Global Thermoelectric is working on planar SOFC operating at 800°C. In July 1997, Global signed a fuel cell agreement with Forschungszentrum Julich, one of the world's leading developers of solid oxide fuel cells. In early 1999 Global reported that they had achieved high levels of power output with a new type of seal and an inexpensive variety of ceramic plates for the stack.

There are few problems with electrolyte management. Liquid electrolytes are usually corrosive and difficult to handle. Solid oxide fuel cells provide some advantages when compared with other fuel cell types. Solid oxide fuel cells are made from solid-state materials, using an ion-conducting oxide ceramic as the electrolyte and operate in the temperature range of 900-1000°C.

A SOFC unit consists of two electrodes (anode and cathode) separated by an electrolyte. The fuel is usually H_2 or CH_4. It is injected at the anode, where it is oxidized by oxygen ions from the electrolyte. This releases electrons (e-) to the external circuit.

On the other end of the fuel cell, oxidant (O_2 or air) is fed to the cathode, where it supplies the oxygen ions (O_2-) for the electrolyte by accepting electrons from the external circuit. The electrolyte conducts these ions between the electrodes.

The current technology employs several ceramic materials for the active SOFC components. The anode is typically constructed from an electronically conducting nickel/yttria-stabilized zirconia cermet (Ni/YSZ). The cathode is based on a mixed conducting perovskite, lanthanum manganate ($LaMnO_3$). Yttria-stabilized zirconia (YSZ) is used for the oxygen ion conducting electrolyte. To generate a suitable voltage, fuel cells are not operated as single units but as a series array of units or stack, with a doped lanthanum chromite ($La0._8Ca0._2CrO_3$) interconnect joining the anodes and cathodes of adjacent units. Several stack designs exist but the most common is the planar or flat-plate configuration. For the YSZ electrolyte to provide sufficient oxygen ion conductivity, a high operating temperature (900-1000°C) is required. This means that expensive high temperature alloys must be used to house the fuel cell.

The cost of the fuel cell could be reduced if the operating temperature were lowered to between 600 and 800°C. This would allow the use of materials such as stainless steel. A lower operating temperature can also reduce the thermal stresses in the active ceramic structures, resulting in a longer lifetime for the system.

To lower the operating temperature, either the conductivity of YSZ must be improved, or alternative electrolytic materials must be developed to replace YSZ. Ceramics that are being investigated include Gd-doped CeO_2, $Ba_2In_2O_5$ and (Sr,Mg)-doped $LaGaO_3$ (LSGM). These materials all face serious drawbacks compared with YSZ, and it is more likely that the first commercial SOFC units will use zirconia-based ceramics as the electrolyte.

Solid Polymer Fuel Cells

The Solid Polymer Fuel Cell (SPFC) is also known as the Proton

Exchange Membrane Fuel Cell (PEMFC). It is unusual in that the electrolyte consists of a layer of solid polymer which allows protons to be transmitted from one face to the other. It essentially requires hydrogen and oxygen as inputs, although the oxidant may also be ambient air. These gases must be humidified.

It operates at a much lower temperature than most fuel cells. The operating temperatures are around 90°C. The SPFC can be contaminated by carbon monoxide, reducing the performance by several percent for contaminant in the fuel in ranges of tens of percent. It requires cooling and management of the exhaust water in order to function properly.

SPFCs are being developed by Ballard, DeNora in Italy, and Siemens. The main focus is transport applications, since there are advantages in having a solid electrolyte for safety. The heat produced by the fuel cell is not adequate for cogeneration. Daimler-Benz is involved in developing cars powered by Ballard fuel cells. Toyota has shown a vehicle that uses a fuel cell of their own design. Other car manufacturers, including General Motors and Ford, are active in similar developments.

The SPFC could be used in small scale power generation, where the heat could be used for hot water or space heating. There is also the potential of a heater/chiller unit for cooling in areas where air conditioning is needed. This particular type of fuel cell could be used for both transport and power generation with the advantages of economies of scale. This could help the introduction of this technology compared to others.

FUEL CELL EVOLUTION

Fuel cells are an old technology, but problems have plagued their utilization. Hydrogen could be widely used but methanol or ethanol are also proposed as fuel as well as gasoline. Fuel cells may be a reality soon in several applications.

Ballard Power Systems is a major developer of fuel cells. In 1993, Ballard's prototype fuel-cell bus was demonstrated and surprised most of those attending an international energy conference. Buses were an ideal platform for a hydrogen-powered fuel cell, particularly for the bulky technology that existed in 1993. They offered a large roof that could be used for fuel storage, a flat floor for batteries and a large engine compartment that can house the cell.

Municipal buses usually run out of a central depot that can be used for hydrogen production and storage. Ballard's bus, like the International Fuel Cells and Daimler-Benz buses that appeared later, proved the concept of drivable prototypes.

Three fuel-cell-powered buses have been in service in Chicago. These buses serve as rolling test beds for Ballard to gather operating data. There is some stack problems, but with forty-five hundred cells in the three buses, there has been only problems in ten of them. The main problems have been with the air conditioning and brake systems rather than the fuel cell drives.

Ballard's buses run on pure hydrogen, without the need for a reformer. The fuel cells develop enough power on their own without the need to be supplemented by battery packs. The fuel cell needs cooling, control, and fuel processing to operate.

The Chicago buses need fast acceleration under 25 miles per hour to merge into traffic. In 1995 Ballard developed a fuel cell stack producing the equivalent of 275-horsepower in a fraction of the space the 125-horsepower 1993 cell required.

In addition to the stack for buses, Ballard is building 50- to 100-kilowatt systems for cars, and a small, under-two-kilowatt portable unit that could power a laptop computer or fit in a soldier's backpack.

Ballard is working on PEFC for transportation and electric powerplants. Most of the PEFC technology is developed in house and they own over 200 patents. They are working with DaimlerChrysler and Ford. According to Merill Lynch, the PEFC fuel cell cars powered by Ballard may be in mass production in 2004.

In 1999 they announced the purchase of a license to DMFC intellectual property from the California Institute of Technology (Caltech) and the University of Southern California (USC). The license is based on technology developed at the Jet Propulsion Laboratory of Caltech and the Loker Hydrocarbon Research Institute at USC.

A separate subsidiary builds stationary 250-kilowatt powerplants to run hospitals and factories, and a much smaller 10-kilowatt model that can power the average home. Ballard builds fuel cells for car manufacturers such as Ford, Volvo, and DaimlerChrysler. A fuel cell car needs to be competitive in price with internal combustion models.

FUEL CELLS FOR TRANSPORT

Fuel cells are being proposed to replace Otto or Diesel engines because they could be reliable, simple, quieter, less polluting and have a greater economy. The internal combustion Otto or Diesel cycle engine has been used in automobiles for over 100 years. It has a life span of about 10,000 hours of operation in automobiles and over 25,000 hours in larger applications such as buses, trucks, ships and locomotives.

Automobile manufacturers have been finding new ways to improve the Otto and Diesel engines. Toyota, for example, has demonstrated an Otto cycle automobile with emissions five times cleaner than present requirements. Volkswagen has a prototype compact four-seater Diesel cycle automobile that gets 100-mpg.

Fuel cells can be considerably quieter than Otto or Diesel cycle powerplants, however fuel cells produce electricity which is not the final form of energy for transportation. The electricity must be converted into mechanical power using an electric motor. The Otto or Diesel cycle produces the required mechanical power directly. Otto and Diesel cycle engines are inexpensive to produce, and use readily available liquid fuels.

Fuel cells using reformers do not produce much less pollution than very advanced Otto and Diesel cycle engines with complex catalytic converters. If vehicles use hydrogen as fuel, a hydrogen supply system would need to be installed.

The DAFC would be simpler than the internal combustion engine, produce greater efficiency and be less polluting. The liquid fuel could be handled by slightly modifying the present distribution equipment. When the DAFC is perfected, it may compete with Otto and Diesel cycle automobiles.

FUEL PROCESSORS

Fuel cells show promise as alternatives to batteries or power generators. The proton exchange membrane (PEM) fuel cell has advantages because of its low operating temperature, high power density, and advanced stage of technical development. However, the fuel used by the

PEM fuel cell is hydrogen, which is not easily transported or stored. In order to take advantage of the existing fuel infrastructure, the PEM fuel cell can be integrated with a fuel processor that converts liquid hydrocarbons into hydrogen. The fuel cell can then use the hydrogen to produce electricity.

A complete fuel cell/fuel processor system could weigh about one kilogram (.45 pounds), much of the weight resulting from fuel storage. It would be fueled by a liquid hydrocarbon such as butane, and could provide 5 watts of base load electric power with 10 watts of peak power for one week. The system could use a compact lithium power battery for load leveling and to meet peak electric power demands.

The technology that makes a compact fuel processor possible is based on the enhanced heat and mass transfer exhibited when fluids flow in and around microstructures. These microstructures consist of machined microchannels up to 500 microns wide and other special structures engineered to enhance chemical reactions or separations. Using many microstructures in parallel, chemical systems can achieve major reductions in size and weight.

The process operations take place in parallel sheets that are machined with many parallel micro-scale features. Combinations of reactor, heat exchange, and control sheets are stacked together to form an integrated system that performs operations such as steam reforming, partial oxidation, water-gas shift reaction, carbon monoxide removal and heat exchange. Each parallel sheet may perform one or more chemical process operations.

One component of the fuel processor, the vaporizer, has been demonstrated at the scale required for a 25-kW fuel cell, using methanol as the liquid hydrocarbon fuel. A device with dimensions of $7 \times 10 \times 2.5$cm vaporized methanol at a rate of 208 mL/minute. Heat was provided by catalytic combustion of a dilute hydrogen stream that would be supplied as the exhaust from the fuel cell anode. The same miniaturization techniques could be used for additional system components such as steam reforming, partial oxidation, water-gas shift, and preferential oxidation reactors.

Using methanol as a fuel means extracting hydrogen. Methanol is sulfur-free and yields hydrogen at 200°C Refining methanol is still a complex process involving many steps, each of which must take place at a specific temperature.

One methanol processor under test provides enough hydrogen to take a vehicle almost 200 kilometers between methanol fill-ups. The range is limited by the size of the fuel tank which is small due to the bulk of the fuel processor.

Another problem is that the fuel processor takes a half-hour to warm up. The processor uses steam to free the hydrogen and it takes this amount of time to get the steam ready.

Another type of fuel processor uses a catalyst, instead of steam, to start the hydrogen production. This system is much smaller and weighs half as much as the steam unit.

Methanol poses a danger as a fuel. It proves fatal if ingested and splashing it on the skin can cause blindness and liver and kidney failure. Since methanol dissolves in water, it can be a threat to underground drinking water supplies. The methanol-based fuel additive MTBE (methyl tertiary butyl ether) is being phased out of gasoline, after the chemical was found in the drinking water of several areas.

HYDROGEN FUEL

One solution is to directly use hydrogen as the fuel. While hydrogen has more energy by weight than any other fuel, about three times more than gasoline, it is hard to get much of this energy in a small fuel tank. A commercially compressed gas tank with hydrogen will take a vehicle about 150 kilometers which is no farther than the best car batteries. Hydrogen is also the smallest of molecules and slips through the smallest holes which is troubling given its natural flammability.

Buses has run a test car 450 kilometers using a liquid hydrogen tank. But the cryogenic technology to store fuel at 200°C (20° above absolute zero) is not mature enough for consumer markets.

There are only a few hydrogen filling stations in the world, but companies such as Texaco Energy Systems which specializes in advanced-fuels are investing in technology to make hydrogen fueling possible. This includes advanced storage tanks. Stronger tanks could compress the hydrogen to greater pressures.

Another technique is to pack tanks full of materials that bind hydrogen, slowing down the molecules without liquefying the gas. Graphite fibers with elaborate nanostructures have been shown to absorb more than 20% hydrogen by weight, allowing more to be squeezed into a tank.

Ballard, DaimlerChrysler and Ford will be testing their technologies in the California Fuel Cell Program. The big appeal of fuel cells is the promise of zero emissions with the potential of a hydrogen-oxygen-water cycle that is sustainable forever.

The Energy Research Corporation is working on large molten carbonate fuel cells for power on ships. One unit uses a diesel reformer for an output of several megawatts in a 15-foot-tall package.

TRANSPORT APPLICATIONS

Automakers are facing tighter regulation of tailpipe emissions and several are investing heavily in fuel cells. DaimlerChrysler, Ford and Ballard Power Systems have spent close to $1 billion on fuel cells and plan to spend at least a billion more to begin mass-producing vehicles.

Japan's four largest automobile makers have invested more than $850 million in fuel cells over the past decade. The internal combustion engine is getting harder to improve and even the most sophisticated designs may have difficulty with newer emissions standards imposed in California and several East Coast states.

Fuel cells are attractive since they free electric cars from battery power. Battery-powered cars are smooth and responsive but these features have been overshadowed by the vehicles' limited range. The fuel cell, unlike batteries, which store a charge, generates electricity. Fuel cells utilize different fuels and materials, but one choice for automotive use is the proton exchange membrane (PEM) fuel cell.

Another way to extend the range of the electric car is to carry fuel and generate electricity on board. This is the approach used by hybrid gasoline-electric cars such as the Toyota Prius.

The Prius uses a small combustion engine, plus a set of batteries to supplement the engine during acceleration. This approach is inherently complicated and costly, since it combines electric and mechanical drive technologies.

Fuel cell vehicle systems are still costly and supplying hydrogen to the unit is a problem. Even compressing the hydrogen at 5,000 pounds per square inch may take up too much space for a 70-mile-per-gallon, 350-mile-range vehicle. Storing the hydrogen in metal hydride is being pursued, but adds weight and high costs. Researchers at Northwestern University have developed a system based on the absorption of carbon nanofilters for the high density of hydrogen. This could make direct-hydrogen cars practical and researchers at the National University of Singapore have reported promising results.

Methanol is easily produced from natural gas or distilled from coal. In the 1980s it was used for internal-combustion engines, but methanol is highly toxic and corrosive. The on-board extraction of hydrogen from gasoline could make the transition to the fuel cell vehicle seamless, but refining gas on the go is difficult. The reactions occur about 800°C, making the devices slow to start, and the chemistry is temperamental. The process is used in chemical manufacturing plants and oil refineries to make industrial volumes of hydrogen.

General Motors and Exxon Mobile recently announced the joint development of a gasoline fuel processor. DaimlerChrysler is developing a methanol system for fuel cells that run directly on methanol rather than hydrogen. (Figure 5-2)

PROTON-EXCHANGE MEMBRANE FUEL CELLS

The proton-exchange membrane fuel cell (PEMFC) is solid, compact and operates at a relatively cool 80°C. The PEM cell uses a rubbery plastic membrane coated with a platinum catalyst. The catalyst splits hydrogen gas into protons and electrons and only the protons can pass through the membrane. The electrons move over the membrane, generating the electric current and then recombine with the protons and oxygen on the other side of the membrane to generate water. A series of the membrane-catalyst assemblies make up a cell. The cells are connected in series to increase the voltage.

PEM cells were used in the Gemini spacecraft in the 1960s, but the amount of power was too low and too expensive to be transferred to commercial applications.

In the late 1980s, Los Alamos National Laboratory made major

Front

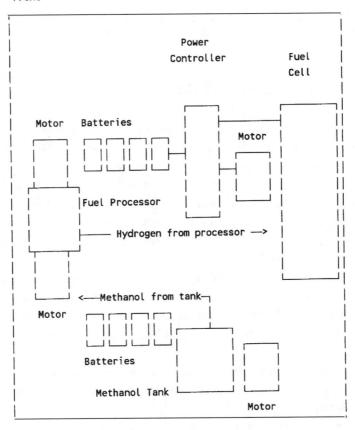

Figure 5-2. A fuel cell car

advances in catalysts, reducing by 90% the amount of platinum required. Ballard Power Systems increased the stack's power density keeping the membranes wet but not soaked and by perfecting the way that hydrogen, oxygen and water move through the stacks. Ballard, a British Colombia based company has almost 400 patents in PEM technology.

A few years ago Ballard exceeded a power density of 1,000 watts per liter. Newer stacks can put out as much as 1,350 watts per liter. This power density should accelerate an automobile as well as an internal combustion engine.

Ballard is commercializing fuel cells in portable power generators, residential generators and stationary powerplants. Three fuels are under consideration: gasoline, methanol and hydrogen.

Many aspects of fuel cell energy involve the relationships that result from thermodynamics, heat transfer, and fluid mechanics.

FUEL CELLS AND THERMODYNAMICS

Thermodynamics involves the relationships of energy transport to properties and characteristics of various types of systems. It allows us to describe the behavior of energy-sensitive devices.

A thermodynamic system is a region with a boundary to define the system. Usually, the system boundary is the physical shell of the unit. A closed system is one where no mass may cross the boundary, while an open system, sometimes called a control volume, will generally have mass flowing through it.

Energy is generally divided into two categories: stored and transient types of energy. The stored forms of energy are potential, kinetic, internal, chemical, and nuclear. Chemical and nuclear energy represent the energy bound up in the structure of the molecular and atomic compounds.

Potential, kinetic, and internal-energy are generally nonchemical and nonnuclear. They relate to the position, velocity, and state of material in a system.

The transformation of a system from one state to another is called a process. A cycle is a set of processes by which a system is returned to its initial state.

A process is said to be reversible if a system can be returned to its initial state along a reversed process line with no change in the surroundings of the system. In practice a truly reversible process is not possible. All processes contain effects that render them irreversible such as friction, nonelastic deformation, turbulence, mixing, and heat transfer loss. The reversible process serves as a reference value. It is the ideal process and represents a theoretical limit.

Many processes can be described by a phase which indicates that one of its properties or characteristics remains constant during the process. These include constant volume (isometric), constant pressure (isobaric) and constant temperature (isothermal).

THERMODYNAMIC LAWS

Thermodynamic laws define the relationships between mass and energy in both open and closed systems. They are based on the conservation of mass for a system with no relativistic effects. In energy conservation, the first and second laws of thermodynamics are the most important.

For steady systems, a steady-state, steady-flow approach may be used. This approach assumes that the state of the material is constant at any point in the system. For transient processes, a uniform flow, uniform state approach can be used. This assumes that at points where mass crosses the system boundary, its state is constant with time. The state of the mass in the system may vary with time but it is uniform at any time.

The first law of thermodynamics implies the conservation-of-energy and gives a balance of energy during a process. The second law includes the prediction of a proposed process or the direction of system changes following a perturbation of the system.

There are two statements of the second law. These can be shown to be equivalent. One of them is the Kelvin-Planck rendition:

It is impossible for any device to operate in a cycle and produce work while exchanging heat only with bodies at a single fixed temperature.

The other statement is the Clausius statement:

It is impossible to construct a device which operates in a cycle and produces no effect other than the transfer of heat from a cooler body to a warmer body.

The utility of the second law includes determining:
- efficiency of heat engines,
- coefficient of performance for heat exchange devices,
- feasibility of a process,
- direction of a chemical or other type of process, and
- physical properties of a system.

EFFICIENCY

A steam engine may be only 30% efficient in converting thermal-energy into external-energy. When converting similar types of energy, the Law of Energy Conservation applies to all processes. Since a larger amount of caloric-energy is turned into a smaller amount of external-energy, a refrigerator can have an efficiency higher than 100%, so a term called the coefficient of performance is used.

Efficiency should be reserved for the conversion efficiency between forms of energy that can be entirely converted into external-energy. These forms are external, kinetic, potential, electrical, mechanical, expansion, Helmholtz, Gibbs, free and exergic energy.

The H-X efficiency can be used to represent the percentage of Helmholtz-energy that can be converted to external-energy. In a large combined cycle gas turbine, the after combustion H-M efficiency can approach 80%. This means that of the Helmholtz energy left in the combustion chamber after the combustion process, 80% can be converted into mechanical-energy.

There is an ideal cycle, called the Carnot cycle, which gives the maximum efficiency for heat engines and refrigerators. It has four ideal reversible processes. The efficiency of the Carnot cycle is:

$$1 - \frac{T_L}{T_H}$$

This represents the best possible performance of cyclic energy conversion devices operating between two temperature extremes, T_H and T_L. The thermodynamic efficiency is different from efficiencies applied to devices that operate along a process line. This efficiency is defined as

$$\frac{\text{actual energy transfer}}{\text{ideal energy transfer}}$$

for work-producing devices and this becomes

$$\frac{\text{ideal energy transfer}}{\text{actual energy transfer}}$$

for work-consuming devices. The actual performance can be calculated from an ideal process line and the efficiency can be determined from the actual energy used.

POWER AND REFRIGERATION CYCLES

Many cycles have been devised to convert heat into work and work into heat. Some of these take advantage of the phase change of the working fluid. These include the Rankine, vapor compression and the absorption cycles. Others contain approximations of thermodynamic processes to mechanical processes and are called air-standard cycles.

The Rankine cycle is used in most large electric generation plants, including gas, coal, oil, and nuclear. It is considered the ideal cycle for a simple steam powerplant. The efficiency depends on the average temperature at which heat is supplied and the average temperature at which heat is rejected. Any changes that increase the average temperature at which heat is rejected or decrease the average temperature at which heat is rejected will increase the Rankine cycle efficiency.

The Rankine cycle has a lower efficiency than a Carnot cycle with the same maximum and minimum temperatures, but the Rankine cycle can be approximated in practice more easily than the Carnot cycle.

Most modern steam-electric powerplants operate at supercritical pressures and temperatures during the boiler heat addition process. This leads to the necessity of reheating between high and low pressure turbines to prevent excess moisture in the latter stages of turbine expansion to prevent blade erosion. Feedwater heating is also used to increase the efficiency of the basic Rankine cycle.

The vapor compression cycle is like a reversed Rankine cycle. The main difference is that an expansion valve is used to reduce the pressure between the condenser and the evaporator. The vapor compression cycle is used for refrigeration and heat pumps.

Air-standard cycles are used for power generation and heating/cooling. They are thermodynamic approximations of the processes occurring in the devices. In the actual cases, a thermodynamic cycle is not completed, requiring the approximations. Air-standard cycles use the following approximations:

1.   Air is the working fluid and behaves as an ideal gas.

2.   Combustion and exhaust processes are replaced by heat exchangers.

COMBUSTION PROCESSES

Natural and manufactured gases, coal and liquid/fuel/air mixtures are energy sources that require combustion. The combustion processes can be analyzed using the combustion equation and the first law of thermodynamics for the combustion chamber. The combustion equation is a mass balance between reactants and products of the chemical reaction combustion process. The first law is the energy balance for the same process using the results of the combustion equation as input.

For hydrocarbon fuels, the combustion equation (chemical balance) is

$$C_xH_y + \alpha(O_2 + 3.76N_2) \rightarrow bCO_2 + c\ CO_2$$

$$+ eH_2O + dO_2 + 3.76\ \alpha N_2$$

This equation neglects the minor components of air. Air is assumed to be 1 mol of O_2 mixed 3.76 mol of N_2. The balance is based on 1 mol of fuel C_xH_y. The unknowns depend on the particular application.

The relative amount of fuel and air are important. The air/fuel ratio controls the temperature of the combustion zone and the energy available to be transferred to a working fluid or converted to work. Stoichiometric air is the quantity of air required when no oxygen will appear in the products.

Ideal combustion means perfect mixing and complete reactions. In this case there is no free oxygen in the products. Most industrial combustion processes conform closely to a steady-state, steady-flow case.

GAS ANALYSIS

During combustion in heaters and boilers, the information required for control of the burner settings is the amount of excess air in the fuel gas. This percentage can be a direct reflection of the efficiency of combustion.

The technique for determining the volumetric makeup of combustion by-products is the Orsat analyzer. The Orsat analysis depends upon the fact that for hydrocarbon combustion the products may contain CO_2, O_2, CO, N_2, and water vapor. If enough excess air is used to obtain complete combustion, no CO will be present. Further, if the water vapor is removed, only CO_2, O_2, N_2 will remain.

The Orsat analyzer uses a sample of fuel gas that is first passed over a desiccant to remove the moisture. The amount of water vapor can be found from the combustion equation.

The sample is exposed to materials that absorb the CO_2, then the O_2, and the CO if present. After each absorption the volumetric change is carefully measured. The remaining gas is assumed to be N_2 but it could contain some trace of gases and pollutants.

FORMS OF ENERGY

In evaluating different types of fuel cells it is useful to consider the different forms of energy. Einstein-energy refers to the total energy in the universe or a particular system. When it is said that energy is conserved in the universe, it is Einstein-energy that is conserved, because other forms may not be.

Einstein-energy can be divided into either external-energy or internal-energy. A major difference between external-energy and internal-energy is the fact that internal-energy can never be completely converted into external-energy.

External-energy is made up of kinetic-energy and potential-energy. Internal-energy includes the following types:

Thermal-energy—The motion or translational energy of molecules.

Chemical-energy—Energy due to the bonding of atoms in molecules.

Radiant-energy—Energy contained in photons.

Substances usually contain mixtures of external-energy and internal-energy. Caloric-energy represents the amount of internal-energy that will flow between two reservoirs. Caloric-energy can never be completely converted into external-energy. It can be split up into two parts for analysis. The Helmholtz-energy is the part of caloric-energy that could be converted into external-energy in a future process. The bound-energy is the energy that could have been converted into external-energy if the conversion had started at an infinite temperature or temperament and progressed until the present point. Gibbs-energy is composed of Helmholtz-energy plus gas expansion-energy. Exergic-energy is composed of external-energy plus Helmholtz-energy.

A fuel cell creates electricity, which is a type of external-energy, directly from the energy in chemical fuels without an intermediate conversion into thermal-energy. When a hydrogen atom bonds to an oxygen molecule, not as much total energy is required in the newly formed water molecule as in the separate hydrogen and oxygen molecules. A certain amount of energy can be released. When the hydrogen-oxygen bonding occurs, the excess energy under ideal conditions can be released in a single packet for each newly created bond. The excess energy is not released in multiple randomly sized amounts of energy. This energy packet is called a virtual photon.

Photons are localized wave packets of energy. When they cannot be detected, they are called virtual. They are referred to as a packet because the energy does not divide and two packages do not join together. This virtual photon can, under ideal conditions, be transferred to other chemical systems, without being lost to the surroundings. Real photons are packages of energy that are found in light.

Joules or Btus could be used as a measure of the amount of energy that each real or virtual photon contains but it would be a very small fraction of a Joule or Btu.

The scale called temperature is used to measure thermal-energy. This represents the average collision energy between molecules. Real photons are created during these collisions.

Radiant-energy is thought of as having a certain temperature. This can be extended and temperature or temperament can be used to represent the amount of energy in all types of photons.

Efficiency describes the effectiveness of devices that operate in cycles or processes. Thermodynamic efficiency is the ratio of the useful work produced to the work required to operate the process. A heat engine produces useful work, while a refrigerator uses work to transfer heat from a cold to a hot region.

REFRIGERATOR ENERGY

A refrigerator creates more caloric-energy than the external-energy input. In a refrigerator more caloric-energy can be created than the external-energy that was put in. By having two heat reservoirs, it is possible to create a potential flow of heat between them which is called caloric-energy. This energy flows from a hot reservoir to a cold ambient reservoir, or from a cold reservoir to a hotter ambient reservoir. If this caloric-energy is created at a very small temperature difference, it does not take much external-energy to create the caloric-energy.

The heat is disposed of by the condenser at the back of the refrigerator at variable temperatures much higher than room temperature and the evaporator draws heat inside the refrigerator at variable temperatures somewhat colder than the refrigerator compartment.

Work is used to represent all forms of external-energy transfer. Heat is used to represent the transfer of thermal-energy. This system has been used for over 150 years, today it is more precise to say that Helmholtz energy was transferred rather than just a simple heat transfer.

In our present system, work is defined in physics as representing all forms of external-energy and internal-energy transfer, but in engineering it represents only the external-energy transfer.

In a gas expansion, there is a type of flow of the multidirectional energy of the molecules from a small space into a larger space. Conversions from one form of external-energy to another form of external-energy such as this can occur in a single reservoir of energy. Converting internal-energy such as thermal-energy to external-energy is more difficult and a flow between two reservoirs is required.

A reservoir is considered to be a source of internal-energy in equilibrium. The flow between two reservoirs can be called caloric-energy. This difference in total internal-energy of two systems results in a flow. All forms of internal-energy can be caloric-energy.

HELMHOLTZ ENERGY

The Carnot-ratio is the amount of Helmholtz energy in an amount of caloric-energy. There is a maximum amount of external-energy that can be extracted from two reservoirs of internal-energy.

In classical thermodynamics the model for thermal-energy is based on the molecular kinetic theory which has molecules colliding and gases expanding. A model based on the quantum physics model has the forces between matter resulting from electromagnetic force fields. When two atoms collide, the impact of the collision is absorbed by the atom's electrons which gain energy. This increased energy results in an unstable orbit for the electron, it wants to shed this energy as quickly as possible. This is because the electron has shifted into an orbit where its vibrations are no longer even divisions of the orbital path. The atom may reabsorb the energy from the electron or a photon is emitted which is radiated away from the atom in the form of radiant energy. The photon is not absorbed into only atoms with zero thermal-energy, it finds any atom to be absorbed into.

The differences in emission versus absorption of the system result in the limited amount of external-energy that can be taken from internal-energy. The caloric-energy that could be potentially converted to external-energy represents the Helmholtz energy. Other forms of internal-energy are due to the bond energy between the different atoms and molecules. Carnot- and Helmholtz-Ratio

The Carnot-ratio indicates the conversion efficiency during a process. The Carnot-ratio (Cr) can be used to represent the external-energy that could be extracted from a flow of caloric-energy at an instant of time. The Helmholtz ratio (Hr) represents the external-energy that could be extracted during the entire flow of caloric-energy.

$$Cr = 1 - T0/T$$

Where:

$T0$ = Constant ambient temperature flow of internal energy, infinitely large reservoir.

T = Constant higher temperature flow of internal-energy.

The Helmholtz energy stored in the system will always equal the amount of external-energy that was added or can be removed. In an irreversible process, there will be a difference in the amount of external-energy added versus the amount of Helmholtz energy stored. In processes where T does not change during the process, the Helmholtz ratio is equal to the Carnot-ratio as shown below:

$$Cr = 1 - T0/T$$

The Carnot-ratio can also be calculated for variable temperature processes of variable specific heat such as steam processes. The Helmholtz-ratio can be calculated for any type of internal-energy. It can be thermal, chemical or nuclear energy.

ENTROPY

In gas compression processes, the amount of external-energy that could be obtained from a flow of caloric-energy is often calculated using a term called entropy. Entropy is not a type of energy. It is a property of a system that is defined by the following differential equation:

$$\delta S = \delta Q/T$$

Where
    S = Entropy of the process
    Q = Total heat transferred
    T = Temperature of the process

Entropy is defined for a reversible process. The third law of thermodynamics states that the entropy of a pure substance >0 at absolute 0 temperature.

Entropy values can be multiplied by the temperature to obtain the bound energy. Entropy is used in processes where the properties of matter are not uniform with temperature and pressure. Tables are used that list entropy at these different conditions. The difference of two values represents the changing entropy during the process.

The Helmholtz function (Hf) is also a property of a system. It is defined as:

$$Hf = U - TS$$

Where:

U =		Total internal-energy
T =		Temperature of the process
S =		Total entropy of the system

In an inefficient process part of the expansion energy is converted into thermal-energy.

If the Helmholtz ratio of a certain system is 0.40 or 40%, it means that a maximum of 40% of the caloric-energy could be converted to external-energy.

Gibbs energy is a mixture of Helmholtz energy which is a type of internal-energy and expansion energy which is a type of external-energy. Since Gibbs energy is a mixture it cannot be used as a pure ratio in the same way as Helmholtz energy.

REVERSIBLE AND IRREVERSIBLE PROCESSES

The multi-directional motion of molecules makes it more difficult for all of the directional energy of external-energy to be converted into external-energy. The concept of entropy represents the disorder. Gas expansion is multi-directional energy and considered to be a form of external-energy. This suggests that some multi-directional energy may be almost entirely converted into external-energy with minimal losses.

Order and disorder may not be as important for systems to be converted entirely to external-energy. This is related to the concept of reversible and irreversible processes. A reversible process is one that can be reversed back to the original conditions by some sort of process and will leave no change in either the system or the surroundings. An irreversible process is one which cannot be reversed back to the original conditions by some sort of process and leave no change in either the system or the surroundings.

External-energy can be stored in many systems that are disorderly, but if there is no change in the system or surroundings, then the system is reversible. This is also true in the case of entropy, since the total entropy of a system is usually unimportant, it is the change in entropy that is important.

A Sterling engine like a Carnot engine can theoretically convert all the Helmholtz energy available in the caloric-energy into external-energy in a system. This engine can also be used in reverse as a heat pump and can convert a quantity of external-energy into caloric-energy.

In an ideal system there is no friction and theoretically no limit to the temperature of operation. If all of the external-energy was converted at an infinitely high temperature and flowed into the Sterling engine system and could be converted without melting the engine, we would be able to extract almost all of the external-energy that was originally transferred into the system. The system would be almost completely reversible. Besides friction that creates the irreversibility, it is the loss of temperature from the maximum that could be achieved.

According to the Energy Conversion Law, in any particular process of converting external-energy to internal-energy or internal-energy to external-energy, any temperature that results that is less than the maximum possible for that process is a loss of external-energy. Internal-energy and external-energy are different entities and yet they must still comply with the Law of Conservation of Energy.

ENERGY TRANSFORMATION

A low amount of caloric-energy can be made into a high amount of caloric-energy in an energy transformer. In an electrical transformer, high voltage, low current power is efficiently converted into lower voltage, higher current power. Internal-energy of a high temperature but low caloric-energy content can also be converted into lower temperature higher caloric-energy content.

One form of internal-energy with a high temperature such as methane fuel could be converted into another type of internal-energy of a lower temperature but higher internal-energy. The internal-energy that is created can be thermal-energy, latent-energy or chemical-energy. In endothermic chemical reactions, some types of energy transformations similar to this do occur.

Most fuel reforming requires a high temperature and the reformation has no ambient body of internal-energy to obtain an additional amount of internal-energy from, so it must draw the internal-energy either from the reaction or from a source of waste heat from a related process. This results in an amount of internal-energy in the final fuel that is not much greater than before. Heat pumps are energy transformers but they require external-energy to power them.

If such an energy transformation could be based on simple chemical reactions, it would be very useful as an energy supply. Many processes such as heating systems in buildings do not require thermal-energy of a high temperature. In an ideal energy transformer, chemical-energy such as found in methane might be converted into a second fuel with several times the original amount of chemical-energy with a much lower temperature. This second fuel could then be burned in a furnace or hot water heater. Less methane fuel would be needed to create the same thermal-energy compared to burning the methane directly in the furnace or hot water heater.

Chemical reactions such as those discussed above or energy conversions such as occur in fuel cells can be theoretically done without any production of thermal-energy. The energy of a chemical reaction can theoretically be entirely converted into other forms of energy such as electricity.

Since the energy is coupled to the field surrounding the atom, there is no leakage. The very high temperature chemical-energy and the low temperature thermal-energy are kept separate.

A heat pump or refrigerator cannot create any more Helmholtz energy in an amount of caloric-energy than the external-energy put in. This Helmholtz energy is the external-energy stored in the heat.

At night, a small amount of thermal-energy may be radiated out from the earth back into space. The average temperature of outer space is estimated to be about 3°K. If it were possible to tap into this cold reservoir, heat engines could be operated very efficiently.

The water stored behind a dam is a form of external-energy. In a hydroelectric dam, the water stored at a higher elevation is a source of potential energy. It is converted to kinetic-energy in the turbines and then to electrical-energy. These are all forms of external-energy. Different forms of external-energy can theoretically be entirely converted to other forms of external-energy.

Generally about 90% of the potential-energy of the water can be converted into electrical-energy which means that the efficiency is 90%. It is also very easy to start and stop any one of the turbines and there is little loss in doing so. Hydroelectric dams are excellent for supplying peaking power during periods of the day when more electric power is used more than average. The potential-energy that is not converted to electric-energy is converted into thermal-energy. This is done at such low temperatures that almost no Helmholtz-energy is left in the thermal-energy.

External-energy can be stored in other devices like flywheels, compressed air tanks, springs, but these devices generally have a much lower energy density than many forms of internal-energy such as chemical-energy.

SOLAR ENERGY CONVERSION

Solar cells presently cannot utilize all the energy of sunlight. Light from the sun has an average temperature of about 6300°K. The Helmholtz ratio of sunlight is about 95%. This means that theoretically it should be possible to convert 95% of the radiant-energy to electricity.

In practice, solar cells may only convert about 10% of the radiant-energy into electricity. This means that 90% of the remaining sunlight is converted into thermal-energy. Solar cells are made of semiconductor material and when the sunlight strikes the semiconductor an electrical potential is created by dislodging electrons. This is due to the impact of the photons. Sunlight is made of photons that contain different amounts of photon-energy at different frequencies. The semiconductor material cannot easily be tuned to convert all types efficiently. This means that some photons will not be converted at all because they have too little photon-energy and some photons will only have a part of their energy converted to electricity because they have too much photon-energy.

Solar cells that have several layers of different semiconductors may be much more efficient. Each layer can be tuned to a specific photon-energy range. Another type of solar cell separates the light into different colors and then each color is converted using a different type of semiconductor. This results in a higher efficiency.

Solar cells may also use lenses or parabolic concentrators to convert more of the radiant energy into thermal-energy. The lens increases the brightness of the light and converts some of the photon-energy into thermal-energy.

A vapor cycle engine could be used to convert this thermal-energy into electricity. This could result in an efficiency of nearly 30% in large installations. One of the advantages with this system is that during the night or on cloudy days fuel can be burned in a separate boiler to operate the system. With the right solar concentrator and engine, efficiencies of over 50% are possible.

A modern gas turbine combined cycle has up to 60% efficiency and uses a much lower temperature of combustion than the sun. The difficulty with solar energy is the path that is required to get the sunlight into the process. Most collection schemes allow radiant-energy to escape. One-way coatings generally only work to keep some of the lower energy rays from escaping but not the higher energy rays.

GAS TURBINE ENERGY

Much of the energy of the fuel in a heat engine is lost in the combustion chamber. Chemical-energy or fuel represents a source of energy that can easily be stored, transported or piped to where it is needed.

Fuel is a type of internal-energy. Engines such as steam engines and diesel engines must first convert the chemical-energy to thermal-energy and then expand the hot gases produced to create external-energy. A typical fuel such as jet fuel can represent a Helmholtz ratio of up to 99%. But, the products of the combustion process tend to dissociate at high temperatures and absorb the energy. This lowers the final temperature produced.

Another problem is that extremely high temperatures begin to oxidize and damage engine parts. Cooling the combustion gases before expansion results in a lower Helmholtz ratio of the combustion gases. In a system like the gas turbine, air is compressed in the compressors, then fuel is added and burned and the hot gases are expanded in the turbine. Jet engines called turbojets use an expansion turbine that is just large enough to compress the incoming air. The thrust is created by the quickly moving hot combustion gases.

Modern turbojet engines use a large turbine which drives a large ducted fan in the front of the engine. Most of the thrust is created by the fan. The same type of gas turbine is used to generate electricity in large generating plants.

Aircraft gas turbines operate with a high pressure cycle. They compress the air to a high pressure before combustion. If the gases in a low pressure ratio turbine are expanded, only a small amount of thermal-energy could be converted into external-energy. In the high pressure gas turbine, the hot gases expand more and more external-energy is extracted.

A gas turbine can be more efficient by operating at a low pressure ratio and using a heat exchanger. In the low pressure gas turbine much of the heat from the hot exhaust gases leaving the turbine can be recycled to the incoming colder compressed air using a heat exchanger. The air coming into the combustion chamber from the heat exchanger can be even hotter than in the high pressure cycle and so less Helmholtz-energy is lost.

The Helmholtz-ratio of natural gas is about 90%. This is the amount of external-energy that is stored in the fuel. Some chemical processes operate at much lower Helmholtz-ratios, some as low as 10%.

The gas turbine has a typical 25-40% loss of Helmholtz-energy in the combustion chamber. These losses are related to combustion irreversibility. Eliminating these losses would increase the efficiency of a gas turbine by 25-40% and result in gas turbines with actual fuel to electricity efficiencies over 80%. There are proposed gas turbines using chemical recuperation and chemical-looping. These may act as energy transformers. In the chemically recuperated gas turbine, the fuel is not turned into thermal-energy in the reformer. The chemical-energy draws in the extra thermal-energy from the gas turbine's exhaust in an endothermic chemical reaction. Theoretically there is no loss of the Helmholtz-energy of the fuel during reforming in this type of reformer that draws in outside thermal-energy. In such a reversible chemical reaction the Helmholtz-energy in the fuel, the Helmholtz-energy in the exhaust and the total caloric-energy are conserved.

In the chemical-looping gas turbine, the fuel first reacts with a chemical substance such as a metal oxide. This reaction is endothermic and the metal pellets that are created have a higher caloric-energy than the fuel had to begin with. The metal pellets are then oxidized with air in the combustion chamber.

In another proposed design, an energy transformer replaces the combustion chamber of a low pressure gas turbine. An ideal Ericsson cycle gas turbine with either a heat exchanger or a bottoming cycle would have a theoretical maximum A-X efficiency of 100% since no Helmholtz energy is lost in the combustion chamber.

An Ericsson cycle heat engine operates similar to a Sterling engine. These heat exchanged engines accept and reject thermal-energy at a constant temperature. While such an engine is not practical, these low pressure gas turbines have a high theoretical efficiency.

ELECTROCHEMICAL FUEL CELL ENERGY

Fuel cells can convert the fuel directly into electricity. A fuel cell creates electrical energy which is a form of external-energy directly from chemical fuels without an intermediate conversion into thermal-energy.

Natural gas fuel would have a Helmholtz-ratio of greater than 0.90 or 90%. This means that theoretically it would be possible to convert 90% of the fuel's internal-energy into electricity. In actual practice there are frictional losses in the fuel cell. In a fuel cell, the Helmholtz-energy in the fuel is converted into ions. These are atoms with less or more of the normal amount of electrons. Frictional losses due to the movement of these ions, and of the electricity that is created, cause significant amounts of thermal-energy to be produced.

Low temperature fuel cells that operate at 80°C and use hydrogen as a fuel can have efficiencies of up to 60%. Higher temperature fuel cells that operate at 1000°K can use hydrocarbon fuels and may reach 80% efficiency by using the thermal-energy produced in a separate heat engine. Since this thermal-energy is created at a high temperature, almost 80% of this thermal-energy produced in the fuel cell from frictional losses remains as Helmholtz-energy.

HYDROGEN FUEL

The hydrogen economy that was promoted in the 1970s was based on producing hydrogen using nuclear powerplants. There would need to a method of making large amounts of hydrogen for a reasonable price. In a hydrogen economy, hydrogen would be used for everything from generating electric power to heating homes and powering industry.

Hydrogen is an extraordinary fuel because only water is produced in operating fuel cells. However, hydrogen is a difficult fuel to store and costly to liquefy. It has a lower energy content than natural gas when pressurized in tanks. There has been some success in storing hydrogen gas in metal hydrides and carbon compounds but most of these techniques require pressure or temperature processing during storage and extraction. Many also require cryogenic refrigeration.

Today, most hydrogen is made from natural gas. Since this process is only 65% efficient when storage losses are considered, this results in a loss of efficiency compared to using the natural gas in a SOFC. Producing hydrogen by electrolysis is generally even less efficient since the electricity generated by a gas turbine is less than 60% efficient.

Hydrogen could be made from the electricity produced by solar panels or fusion powerplants, but the cost of making hydrogen from the electricity of solar panels is much higher than making it from natural gas. Electricity is presently sold for about 3 times the cost of the fuel which makes selling electricity more practical than producing hydrogen.

Ethanol may be the preferred fuel for portable fuel cells. Methanol and ethanol can be made from either natural gas or biomass. This process is about 65% efficient.

Hydrogen and alcohol cost about the same to produce and store. A DMFC is slightly less efficient than a PEFC operating on stored hydrogen gas.

In the future it may also be possible to produce alcohol directly in solar panels or in fusion powerplants. It is also possible to extract carbon dioxide from the atmosphere in the same way as plants do. Artificial photosynthesis using a type of solar panel is being worked on in Japan. If this is practical, carbon dioxide would just be recycled after it is emitted.

To change the trickle of hydrogen produced now for industrial use into a major fuel could cost hundreds of billions. Methanol fuel could bridge the gap for the decade or more it might take to build that infrastructure. Another possibility is service stations with electrolysis plants that produce hydrogen from water. Arete Corporation projects that photovoltaic cells will be used on the roofs of service stations to produce power to make hydrogen locally.

A reformer is like a miniature refinery. Reformed hydrogen is not pure and is not likely to deliver the same performance as hydrogen gas.

Gasoline-reformed fuel cells avoid the hydrogen safety problem. Ever since the 1937 fire that killed 36 people and destroyed the German zeppelin Hindenburg in Lakehurst, New Jersey, airships using hydrogen have not flown.

Hydrogen filled sixteen cells in the airship's body, but highly flammable cellulose doping compound was used to coat the fabric covering, and lacquer, another flammable substance, was used to coat the support structure.

An electrical discharge probably ignited the airship's skin, which was coated with aluminum power. As it docked, the heat from the fire exploded the hydrogen cells and caused the airship to rise and burn.

Some companies that make fuel cells believe that hydrogen is too volatile and dangerous. Hydrogen is the smallest molecule and leaks out of everything. You cannot see it burn.

In our gasoline economy, 15,000 cars are destroyed by engine fires every year, and 500 people die from auto accident-related burns. Gasoline is also highly volatile, but today, we are so familiar with gasoline that it no longer seems very dangerous.

Hydrogen, when spilled, escapes upward instead of puddling like gasoline. It is odorless and its flame is invisible. It emits very little radiant heat, and standing next to a hydrogen fire you might not even be aware of it. Even in diluted form, hydrogen will burn easily, but unless you are in physical contact with the fire, it will not hurt you.

One safety advantage of fuel cells is that they do not burn their fuel, making ignition less likely in the event of a collision or leak. Hydrogen is safer to carry around than gasoline. If we had a hydrogen economy and someone proposed introducing gasoline, it would probably be prohibited as too dangerous. Hydrogen is 52 times more buoyant, and 13 times more diffusive, than gasoline. Victims usually survive better in a hydrogen fire. They are not burned unless they are in it.

Hydrogen tanks for cars may be made of reinforced composites with an aluminum liner. Tank makers have tested their products with 50 foot drops have remained intact, without leaks. Additional strength may be needed if internal tank pressures, now at 3,000 pounds per square inch, are increased to 5,000 pounds. Another problem, which can be addressed with venting and gas-detecting sensors, is the trend for hydrogen to accumulate in eaves of buildings and structures.

FUEL CELLS AND HEAT ENGINES

Heat engines are theoretically at a disadvantage compared to fuel cells. The virtual photons that are transferred during chemical reactions in a fuel cell have an energy somewhere between 3,500° and 20,000° Kelvin. It is this high energy level that allows the fuel cell to be theoretically so efficient. Carnot's Law is usually applied only to the amount of external-energy that can be extracted from thermal-energy systems. The same law should apply to all internal-energy systems. The amount of external-energy that can be extracted from all types of internal-energy is the Carnot-ratio. The Carnot-ratio for virtual photons of 3,500°K is about 90%. This is much higher than for real photons in a gas turbine with an energy level of 1000°K and a Carnot-ratio of 70%. The Carnot-ratio is based on the ambient temperature of the surroundings. The Carnot-ratio applies to the absolute temperature scale where 0°C = 273.15°K. Heat engines such as gas turbines can be considered inferior to fuel cells since they must convert the high level chemical-energy into low level thermal-energy first. A gas turbine cannot operate at the high level of the chemical-energy without melting. As the energy level is reduced, the Carnot-ratio is reduced. A large part of the Helmholtz-energy that was available at the higher energy level is lost, it is converted into bound-energy.

The fuel cell gets warm because of the resistance and inefficiencies from the ion and electron flow during the production of electricity. So, many types of fuel cells can run efficiently at low temperatures while at the same time converting very high level energy.

Advanced gas turbines can achieve a temperature of about 1150°K or 877°C. These gas turbines are used with heat exchanging or steam turbines and can be almost 60% efficient in converting fuel to electricity. In the future, ceramic gas turbines could reach 70% efficiency. This could result in a higher efficiency than a fuel cell can achieve by itself. The possible use of energy transformers in the combustion process could increase the efficiency to 80%.

A second-law analysis of fuel cells can be made for 30-kW powerplants operating on hydrocarbon fuel. The fuel cell process is divided into six subsystems. In each subsystem there are inefficiencies involved that reduce the energy that is left in the system. It shows that the SOFC 30-kW system will have an efficiency of 1.4 times that of the PEFC and 1.3 times that of the DMFC.

The SOFC is the most efficient largely because of the low reformer and air pressurization losses. This is because the SOFC can reform fuel inside the stack and utilize some of the stack waste thermal-energy. Since the PEFC operates at a lower temperature this is not possible. The SOFC does not need to operate at higher than ambient air pressure. It uses a low pressure blower to drive air through the cell. The PEFC runs at a high air pressure. In a small 30-kW powerplant this pressure-energy cannot be readily recovered. The DMFC stack efficiency is very low, but because there are no reformer losses and less air pressurization and system losses, the final efficiency is still higher than the PEFC.

The PEFC operating at ambient air pressure and using hydrogen as it's fuel would be the most efficient fuel cell without using a bottoming cycle such as a gas turbine. It could achieve almost 60% efficiency, while the SOFC would be closer to 50% and the DMFC would be closer to 40%.

Fuel cells are just a few years away from commercialization on a large scale. The fuel cell and technology that will dominate depends on how fast some of the existing problems are solved. The SOFC and the DAFC have problems, but if these can be solved quickly then these may become the predominant fuel cells in the future. There has been considerable progress made in this direction. Ballard has purchased the DAFC technology from JPL, and Global has made progress towards commercializing the SOFC technology. JPL predicts that direct oxidation, liquid feed methanol fuel cell efficiency will increase to 45% with the use of advanced materials.

Among the different types of fuel cells, the proton-exchange-membrane (PEM) cell, is considered one of the best for transportation applications. The PEM cell was developed for the Gemini space program by General Electric in the early 1960s. It has advantages in size, low operating temperatures, adjustable power outputs and quick starting.

Several breakthroughs occurred at the Los Alamos National Laboratories in the 1980s including the reduction of up to 90% the amount of the expensive catalyst needed to coat the cell's thin polymer membrane.

Fuel cells used as electric powerplants may be successful before vehicular applications. A simple type of solid oxide fuel cell (SOFC) would be suitable for 1-30-kW powerplants.

The solid oxide fuel cell is considered to be the most desirable fuel cell for generating electricity from hydrocarbon fuels. This is because it is simple, highly efficient, tolerates impurities and has some capability to internally reform hydrocarbon fuels.

One advantage of the SOFC over the MCFC is that the electrolyte is a solid. This means that no pumps are required to circulate the hot electrolyte. Small planar SOFCs of 1-kW made with very thin sheets could result in a very compact package.

Another advantage of the SOFC is that both hydrogen and carbon monoxide are used in the cell. In the PEFC the carbon monoxide is a poison, while in the SOFC it is a fuel. This also means that the SOFC can use many common hydrocarbons fuels such as natural gas, diesel, gasoline, alcohol and coal gas.

In the PEFC an external reformer is required to produce hydrogen gas while the SOFC can reform these fuels into hydrogen and carbon monoxide inside the cell. This results in some of the high temperature thermal-energy that is normally wasted being recycled back into the fuel.

Since the chemical reactions in the SOFC occur readily at the high operating temperatures, air compression is not needed. This results in a simpler, quieter system with high efficiencies. Exotic catalysts are not necessary.

Many fuel cells such as the PEFC require a liquid cooling system but the SOFC does not. Insulation is used to maintain the cell temperature on small systems. The cell is cooled internally by the reforming action of the fuel and by the cooler outside air that is drawn into the fuel cell.

Since the SOFC does not produce any power below 650°C, a few minutes of warm-up time is required. As an automotive power source, this time period is considered to be a disadvantage but since electric powerplants run continuously, this time period is not a problem. The SOFC may well be suited to certain vehicles which run more continuously such as buses or trucks.

Because of the high temperatures of the SOFC, they may not be utilized for sizes much below 1,000 watts or for small to midsize portable applications. Small SOFCs are about 50% efficient at about 15%-100% power. To achieve even greater efficiency, medium sized and larger SOFCs are generally combined with gas turbines. The fuel cells are pressurized and the gas turbine produces electricity from the extra waste thermal-energy produced by the fuel cell. The resulting efficiency of this type of SOFC generating system is 60 to 70%.

A SOFC suitable for producing 1-30-kW and using natural gas as it's fuel would use a reforming chamber. On the anode side, natural gas is sent into the reforming chamber where it draws waste thermal-energy from the stack and is converted into hydrogen and carbon monoxide. It then flows into the anode manifold where most of the hydrogen and carbon monoxide is oxidized into water and carbon dioxide. Part of this gas stream is recycled to the reforming chamber where the water is used in the reformer.

On the cathode side, air is forced into the heat exchanger where it almost reaches the operating temperature. The air is brought up to the operating temperature of 800°C by combustion of the remaining hydrogen and carbon monoxide gas from the anode. The oxygen in the cathode manifold is converted into an oxygen ion which travels back to the anode.

FUELS

A major question is the fuel itself. Will fuel cells run on pure hydrogen and have a high-compression tank of this highly flammable gas nearby or will they use a reformer to extract hydrogen from a fossil fuel such as methanol? The direct hydrogen approach is cleaner but autos will probably retain their familiar liquid fuels and the first fuel-cell cars will probably run on them.

In 1997, a joint project of Arthur D. Little, Plug Power and the Department of Energy demonstrated a gasoline reformer. This was a major achievement, since gasoline is one of the hardest fuels to reform. Gasoline contains sulfur, which poisons fuel cells, but Epyx which is part of Arthur D. Little trapped the sulfur before it got to the cell using a technique similar to a catalytic conversion. Such a reformer could work with multiple fuels and be changed to use gasoline, ethanol, or methanol.

Chrysler who had been a proponent of gasoline reforming has switched to methanol. This is a result of it's merger with Daimler-Benz. As clean as they are, fuel-cell cars with reformers still are not zero-emission vehicles, as defined by California standards.

References

Boyle, Godrey, Editor, *Renewable Energy Power for a Sustainable Future*, Oxford, England: Oxford University Press, 1996.

Fairley, Peter, "Filler Up with Hydrogen," *Technology Review*, Vol. 103 No. 6, November-December, 2000, pp. 56-60.

Motavalli, Jim, *Forward Drive*, San Francisco, CA: Sierra Club Books, 2000.

Turner, Wayne C., *Energy Management Handbook*, The Fairmont Press, Inc.: Lilburn, GA, 1997.

Internet: www.benwiens.com/energy/"Energy Science Made Simple," September 22, 1999.

Internet: www.benwiens.com "The Future of Fuel Cells," March 19, 2001

Internet: www.pnl.gov "Small Fuel Processors Power Light-Weight Solders' System," April 18, 2001.

CHAPTER 6

MODULAR POWER GENERATION—
TURBINES, GASIFICATION,
COMBINED CYCLE GENERATION

MODULAR POWER GENERATION

There will be new applications of cogeneration in the near future. During most of the 1980s, there was a reprieve from the energy problems of the middle and late 1970s. Now there is uncertainty over electric power supplies.

There will be a large need for innovative new energy systems. It is expected that cogeneration systems will play a large role on residential, commercial, and industrial levels. Innovation will be the key. New types of systems that are both technologically and economically feasible will be developed since the technology exists for both large and small cogeneration systems.

Cogeneration is the process of producing and utilizing two types of energy at the same time. This usually means that the generating system has one primary type of energy it is producing and which will be used and another type that is normally wasted. The cogeneration system captures the waste energy and converts it into a usable form. Electrical generating facilities produce two types of energy, electricity and heat. A cogeneration facility will use both types of energy. Typically, in the electrical generating system, the heat is released into the air or cooling water. Cogeneration is the process by which both types of energy are utilized. See Figure 6-1.

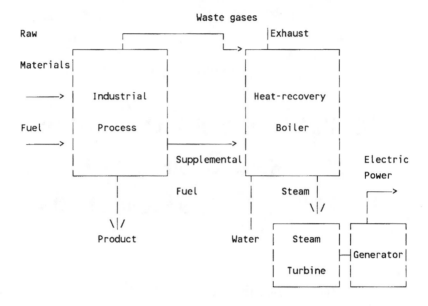

Figure 6-1. Steam cogeneration system

Cogeneration systems have been used mostly for large industrial systems. This has been due to the ratio of initial cost to payback. Cogeneration equipment is relatively expensive both to purchase and to install. If there is not enough energy to recover the cost, the recovered energy may be too high.

However, cogeneration systems can be used anywhere that energy is wasted. There are usable amounts of heat expended by a number of household appliances. These include air conditioning compressors, refrigerators, freezers and clothes dryers.

Recent residential cogeneration systems include a gasoline engine powered electrical generator with a heat exchanger. The heat exchanger transfers heat from the engine to a closed water circulation system to provide hot water. The unit is heavily insulated to trap both heat or noise. Another system uses a heat exchanger to capture the waste heat from central air conditioning units. This system also uses the recaptured heat to fill water heating needs.

Cogeneration systems have been used for decades and have gone through a number of cycles. In each cycle, different types of technologies have evolved to satisfy the market.

In the early years of electric power, cogeneration systems were used because of the unreliability of utility supplied electrical power. Businesses needed their own generating systems if they were to have power on demand and since they were already making an investment in generating systems, they could spend a little more and get a cogeneration system that had greater efficiency.

Once utilities began to become more reliable and less expensive, the popularity of generating and cogenerating with individually owned systems began to decline. These individual generating systems had only a specialized small market and were seldom used. The next cycle occurred during the energy shocks of the 1970s when cogeneration became a more common type of system.

We are now in another cycle where an increasing number of cogenerating systems are being designed and installed. It also appears that the popularity of cogeneration systems may not go through the wide popularity swings as it has in the past. This is because of the underlying changes in the modern economy. Instead of being in an industry-centered economy, which means concentration, there is the shift to a technology-based economy, which allows dispersion. This is coupled with a heightened concern for efficiency and the environment, provides basic strength for the cogeneration market.

The cost savings for cogeneration systems can be substantial, since they typically harness energy that is already being produced by some other process and is being wasted. The cogeneration process is using free energy that is doing nothing to generate electricity.

In order to keep the cogenerator operating efficiently, it must run for extended periods of time, so there may be times when the local demand for electricity will be less than the available power. In these circumstances, it is desirable to sell power to the utility company.

Prior to about 1980 very few utilities would purchase excess power from a customer. In 1978, Congress passed legislation which led to electric utilities being required to purchase cogenerated electricity. It specified that the rate at which the utility pays for electricity must be based upon the avoided cost. These costs are determined from the value of the fuel which the utility would otherwise have purchased to meet the system load, as well as savings associated with not having to expand their facilities or purchase power from other utilities in order to meet the demand.

TOPPING AND BOTTOMING SYSTEMS

Cogeneration systems can be classified as topping or bottoming systems. A topping system is used for generating electricity and a by-product of generating electricity, the heat, is used for some other use. A topping system is an electric generator with heat exchangers to provide hot water for the facility.

A bottoming system has some other energy process take place first, and the waste energy from that source is used to produce electricity (Figure 6-2). This is common in industrial applications where high temperature or high pressure steam is produced and normally wasted after the manufacturing processes are completed.

Another by-product of industrial processes are the gases that are emitted. Rather than emitting them directly into the atmosphere, they can be piped into a turbine which turns a generator to produce electricity.

Although the initial cost of this type of system can be high, the payback can occur from 2 to 7 years where the system will have paid for itself, and begin to provide energy savings. The rate of payback depends on the efficiency of the system and by the cost of buying utility supplied electricity.

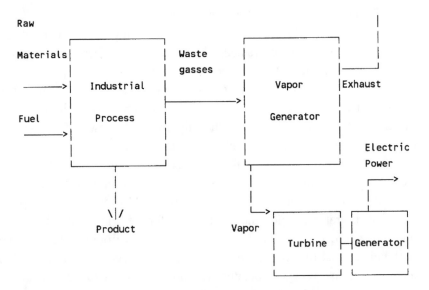

Figure 6-2. Cogeneration bottoming system

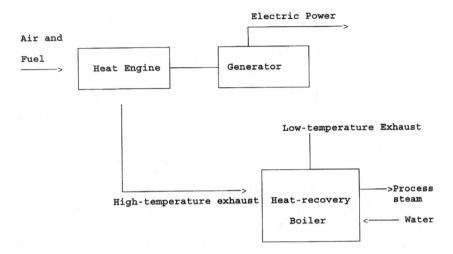

Figure 6-3. Topping system

In a topping system, a heat engine generator is used to provide electricity (Figure 6-3). Then the heat produced by the engine is also utilized. The exhaust gases are piped into a heat recovery boiler which produces steam for industrial processes.

Cogeneration equipment falls under Article 445 of the National Electrical Code (Generators). The provisions of Article 445, along with other applicable parts of the code must be complied with for residential, commercial, or industrial cogeneration systems. This covers only the electrical portion of the cogeneration system. Cogeneration systems also have thermal and mechanical systems which must meet other applicable codes.

PEAK REDUCTION

One use of generators and cogeneration is peak reduction or shaving. The reduction of peak loads can result in significant savings. Utility companies may charge industrial customers a separate demand charge, in addition to their regular electric rates. These charges can be substantial and if the customer can avoid these charges, notable amounts may be saved.

Since the utility company must provide power upon demand, in the amounts demanded, their transmission wires, poles, and transformers must be sized according to the highest possible demand that will be placed upon them. A facility with heavy equipment can have high momentary demands for power. Large electric motors often have a starting current of eight times their normal operating current. For industrial customers, utility companies have needed to install demand meters, and to charge high rates for maximum demands.

The demand charge is actually a reservation fee paid by the customer regardless of the use of the standby service. Ratchets are also used which require the customer to pay the reservation fee for an extended period. These requirements vary greatly from state to state.

Beyond larger power lines and transformers, the utility company has to build more or generating capacity because of these maximum demands. Because of this, some electric utilities offer large customers economic incentives if they enroll in load curtailment programs.

These are programs in which the utility and the customer cooperate on a contractual basis to reduce customer demand during periods of high power usage, such as in the summer months when air conditioner usage is high.

A typical load curtailment contract may specify that the utility may at its option and within a specified time period, usually between 30 minutes and 2 hours, request the customer to cut back or curtail the load. These programs are helpful to the utility, since they can substantially reduce total demand during periods of peak usage. Peak demand drops of 1000-kW could save a customer over $100,000 per year.

Originally, standby service was designed to provide power to customers which have their own generation during outages. Today, standby service means that customers can seek supplies on the open market, protecting themselves from interruptions.

Standby rates were designed in the early years of power generation to discourage interconnection by those that generated their own electricity. Utilities priced the service at a level that made it uneconomical for customers to implement on-site generation. The Public Utility Regulatory Policies Act (PURPA) of 1978 made this illegal and directed the states to implement reasonably priced standby service.

EXCESS POWER

The electricity sold to the utility company must be of the same frequency and voltage characteristics as the utility company's power. The frequency of the generated electricity and the frequency of the power supplied by the utility must also be synchronized.

This is commonly done for large installations by synchronizing controllers. These controls operate the synchronous generator (alternator), monitor the utility company's power and synchronize the two systems. Synchronous generators are like the synchronous motors used in clocks, they must run in step with the electric system frequency. They are more often referred to as alternators and are used in 1000-MW steam turbine generating plants. Small alternators are also used in wind energy systems of 1 to 10-kW. Most of these smaller units are permanent magnet types.

Alternators can generate electricity at unity power factor and because they are self exited they can operate independently of the central power station. Some alternators operate at their own speed and frequency. This variable frequency output is used for resistance heating, welding or rectified to charge batteries. It can also be fed to a static converter for conversion to 60-Hz for parallel operation with utility power.

DC generators are used in battery systems or for loads such as heating. They are also used with static inverters to power AC loads or send power to the power grid. DC generators include rotating motor-generators along with photovoltaic or electrochemical fuel cell sources.

There is also the option of using an induction generator which will supply power with the same characteristics as the utility company's power. Induction generators are self-regulating, adjusting to varying torques and load conditions.

An induction generator is basically an AC squirrel-cage motor. When connected to the utility company's power, it will run at its standard speed, which is usually about 4% below the speed of the rotating magnetic field within the motor, called synchronous speed.

While the motor is running on the utility company's power, if you drive the motor faster than the synchronous speed, the motor will generate power back into the utility company's lines. When operating in this mode, it acts as an induction generator since it is no longer working as a motor but as a generator.

When connected this way, the induction generator will generate electricity with the same exact characteristics as the utility power. No synchronizing controller is required and driving the machine faster will not change the frequency of the power it produces, it just increases the amount of power produced. Induction generators cease generating when the utility source is cut off, unless power factor capacitors are connected.

POWER FACTOR

One disadvantage of induction generators is the power factor that worsens when the kW output is reduced. The power factor is a ratio between the true or real power and apparent power. The highly magnetic loads of motors and transformers increase the apparent power which does not show up on standard power meters.

For industrial customers, utility companies install power factor meters, and charge special power factor charges, which can be costly. A large effort goes into the reduction of power factors. This is usually done by adding capacitors, since their capacitive reactance offsets the inductive reactance from magnetic loads.

If the generator goes off line, there must be a time delay of several seconds before it can be reconnected. If it were immediately reconnected, there would be large transients due to the phase differences between the decaying generator voltages and the utility voltage. A waiting period of several seconds is usually sufficient.

A controller is also required to disconnect the motor/generator from the power lines if utility power drops out. This is called a no voltage feature and often involves a special no voltage relay. Low voltage units, as opposed to no voltage units are also available.

CONVERTERS

Modern converters are solid state electronic units that convert AC to DC, DC to AC or AC to AC at another voltage and frequency. They are usually called inverters when used for DC to AC conversion. Line commutated converters depend on an AC source, usually grid power for their operation. They require lagging, reactive loads which may equal or be greater than the converter power rating. The chopped wave output of these converters can cause metering errors, electrical interference and excessive heating in motors.

Self commutated converters are more expensive, do not require central station power and can operate by themselves to supply standby power. They generally supply a better waveform. A synchronous converter may be a line commutated or self commutated device which is triggered by the utility and synchronizes its output with the utility.

QUALITY OF VOLTAGE

Problems with flicker and with alternately bright and dimming lights are usually associated with varying loads and motors starting currents on the same power circuit. They may also be caused by generators with varying output. Voltage problems are related directly to the size of the generator as well as the capacity of the circuit to which it is connected.

The likelihood of problems from a single cogenerator can be minimized if the rated output is limited to about 1/2 of the distribution transformer rating. This means about two to ten kW for a single phase unit or 25 to 100-kW for three phase units. The larger values are used for locations near the substation.

Other problems with voltage control can appear as additional cogenerators are connected. Especially if the capacity of on-site generators becomes significant compared with the total load. A wider range of regulation may be obtained with additional line regulators.

Electric system operating conditions may vary from minimum load with all generators operating to full load with all generators off. The connection might need to include the increased circuit capacity or a separate circuit for the power producers.

TURBINES

Small turbines are becoming important as auxiliary sources of power. The gas turbine and the steam turbine were conceived simultaneously. In 1791, John Barber's patent for the steam turbine described other fluids or gases as potential energy sources. Barber's gas turbine was a unit in which gas was produced from heated coal, mixed with air, compressed and then burnt. This produced a high speed jet upon the radial blades of the turbine wheel rim. Others before him that recorded similar schemes, include:

- Giovanni Branca - impulse steam turbine—1629,
- Leonardo da Vinci - smoke mill—1550,
- Hero of Alexandria - reaction steam turbine—130 BC.

These early gas turbines schemes would today be more accurately called turboexpanders, since the source of compressed air or gas is a by-product of a separate process. These ideas were not turned into working equipment until the late 19th Century when Charles de Laval and others produced working hardware. The use of steam turbines grew and the technology became available to gas turbines, gas generator compressors and power-extraction turbines.

The axial flow compressors of today's gas turbines resemble the reaction steam turbine with the flow direction reversed. The similarities between steam and gas turbine components are rooted in their common history.

In 1905, a gas turbine and compressor unit was installed at the Marcus Hook Refinery of the Sun Oil Company near Philadelphia, PA. It provided 5,300 kilowatts (4,400 kilowatts for hot pressurized gas and 900 kilowatts for electricity). The first electricity generating turbine for a power station was built at Neuchatel in Switzerland in 1939. This 4,000-kilowatt turbine used an axial flow compressor delivering excess air at 50 pounds per square inch to a single combustion chamber and driving a multi-stage reaction turbine. Excess air was used to cool the exterior of the combustor and to heat that air for use in the turbine.

An early utility gas turbine powerplant in the U.S. was the Huey Station unit of the Oklahoma Gas & Electric Company in Oklahoma City. This 3,500-kilowatt unit went on-line in 1949. It was a simple-cycle gas turbine with a fifteen stage axial compressor, six straight flow-through combustors placed circumferentially around the unit, and a two-stage turbine.

During World War I, the reciprocating gasoline engine was being refined for the small, light aircraft of the time. The gas turbine was big and bulky, with too large a weight-to-horsepower output ratio to be considered for an aircraft powerplant. However, the turbo-charger became an addition to the aircraft piston engine. The exhaust-driven turbo-charger was developed in 1921, which led to the use of turbo-charged piston engine aircraft in World War II.

In 1937 British Thomson-Houston Company built and tested Frank Whittle's jet engine. It consisted of a double entry centrifugal compressor and a single stage axial turbine. A turbojet engine consisting of a compound axial-centrifugal compressor similar to Whittle's design and a radial turbine was built by the German aircraft manufacturer Heinkel. In 1939 a turbojet aircraft powered by this engine made the first flight of a jet powered aircraft.

Throughout the war years various changes were made in the design of these engines. Radial and axial turbines were used along with straight through and reverse flow combustion chambers, and axial compressors. The compressor pressure ratio started at 2.5:1 in 1900, went to 5:1 in 1940, 15:1 in 1960, and is currently approaching 40:1.

Since World War II, improvements made in aircraft gas turbine-jet engines have been transferred to stationary gas turbines. Following the Korean War, Pratt & Whitney Aircraft provided the cross-over from the aircraft gas turbine to the stationary gas turbine. In 1959, Copper Bessemer installed the world's first aircraft industrial gas turbine, in a compressor drive. This unit generated 10,500 brake horsepower (BHP) driving a pipeline compressor.

In airborne applications the units are referred to as jets, turbojets, turbofans, and turboprops. In land and sea-based applications the units are referred to as mechanical drive gas turbines.

Jet engines are gas generators where the hot gases are expanded either through a turbine to generate shaft power or through a nozzle to create thrust. Some gas generators expand the hot gases only through a nozzle to produce thrust. These units are identified as jet engines or turbojets. Other gas turbines expand some of the hot gas through a nozzle to create thrust and the rest of the gas is expanded through a turbine to drive a fan. These units are called turbofans. When a unit expands most of its hot gases through the turbine driving the compressor, and the attached propeller and no thrust is created from the gas exiting the exhaust nozzle, it is called a turboprop. Turboprops have much in common with land and sea-based gas turbines. The engines used in aircraft applications may be either turbojets, turbofans, or turboprops, but they are commonly called jet engines.

The turbojet is the simplest form of gas turbine since the hot gases generated in the combustion process escape through an exhaust nozzle to produce thrust. Jet propulsion is the most common use of the turbojet, but it has been adapted to drying applications, supersonic wind tunnels, and as the energy source in a gas laser. The turbofan combines the thrust provided by expanding the hot gases through a nozzle (as in the turbojet) with the thrust provided by the fan. The fan acts as a ducted propeller. In recent turbofan designs the turbofan approaches the turboprop in that all the gas energy is converted to shaft power to drive the ducted fan. Turboprops use the gas turbine to generate the shaft power to drive the propeller, there is no thrust from the exhaust.

In the 1967 Indianapolis 500 Race a Pratt & Whitney turboprop powered car led the race for 171 laps, only to have a gearbox failure on the 197th lap. The car had an air inlet area of 21.9 square inches. Later, race officials modified the rules by restricting the air inlet area to 15.999 square inches or less. A year later race officials further restricted the air inlet area to 12.99 square inches. This effectively eliminated gas turbines from racing.

While some engines are derivatives of these aircraft engines, a majority of land based gas turbines were derived from the steam turbine. Like the steam turbines, these gas turbines have large, heavy, horizon-tally split cases and operate at lower speeds and higher mass flows than the aircraft derivatives at equivalent horsepower. A number of hybrid gas turbines in the small and intermediate size horsepower range have been developed to incorporate features of the aircraft derivatives and the heavy industrial gas turbines.

In the mid-1960s, the U.S. Navy implemented a program to use gas turbines as a ship's propulsion powerplant. The first combat ship con-structed was the *USS Achville*, a Patrol Gunboat which was commis-sioned in 1964. The U.S. Navy has also outfitted larger ships. The Arleigh Burke Class Destroyer used four aircraft derived gas turbines as the main propulsion units with 100,000 shaft horsepower. By the end of 1990s, the U.S. Navy had over 140 gas turbine propelled and 27 navies of the world had over 330 ships with some 800 gas turbines.

Gas turbines have also been used to power automobiles, trains, and tanks. The Abrams tank is equipped with a gas turbine engine. This 63 ton unit can travel over 40 miles per hour on level ground.

Gas turbines can have many different forms, single or dual shaft, hot or cold end drive. A gas turbine can be viewed as a gas generator and a power-extraction-turbine, where the gas generator consists of a compressor, combustor, and compressor-turbine (Figure 6-4). The compressor-turbine is the part of the gas generator that develops the shaft horsepower to drive the compressor. The power-extraction-turbine is the part of the gas turbine that develops the horsepower to drive the external load. The energy that is developed in the combustor, by burning fuel under pressure, is the gas horsepower (GHP).

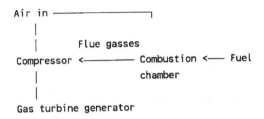

Figure 6-4. Conventional gas turbine

On turbojets, the gas horsepower that is not used by the compressor-turbine to drive the compressor is converted to thrust. On turboprops, mechanical drive, and generator drive gas turbines the gas horsepower is used by the power extraction-turbine to drive the external load.

The gas horsepower may be expanded through the remaining turbine stages, as done on a single shaft machine, or through a free power turbine, as done on a split shaft machine. The additional energy is converted into shaft horsepower and depends on the efficiency of the power extraction turbine.

A single spool-split output shaft gas turbine, also called a split-shaft mechanical drive gas turbine, is a single-spool gas turbine driving a free power turbine. The compressor/turbine component shaft is not physically connected to the power output (power turbine) shaft, but is coupled aerodynamically. This aerodynamic coupling, also called a liquid coupling, allows easier, cooler starts on the turbine components. It allows the gas turbine to reach self-sustaining operation before it drives the load. The gas turbine can operate at this low idle speed without the driven equipment rotating. This type of configuration is used in compressor and pump drives as well as electric generator drives.

This arrangement also allows the power turbine to operate at the same speed as the driven equipment. In generator drive applications the power turbines may operate at either 3,000 or 6,000 rpm to match 50-cycle or 60-cycle generators. Centrifugal compressor and pump application speeds are usually in the 4,000 to 6,000 rpm range. Matching the speeds of the drive and driven equipment eliminates the need for a gearbox.

One new power source is the 20 to 60 kilowatt, regenerated, gas turbine power package. This package, in combination with a battery pack, can deliver low emission power in automobiles.

TURBINE EVOLUTION

The growth of the gas turbine in recent years has been driven by metallurgical advances that allow high temperatures in the combustor and turbine components. Other factors include both aerodynamic and thermodynamic breakthroughs and the use of computer technology in the design and simulation of turbine airfoils and combustor and turbine blades. There have been improvements in compressor design, increases in pressure ratio, combustor and turbine design.

Gas turbines have always been tolerant of a wide range of fuels from liquids to gases, to high and low Btu heating values and are now functioning satisfactorily on gasified coal and wood.

Another factor has been the ability to simplify the control of a highly responsive unit using computer control technology. Computers start, stop, and govern the operation of the gas turbine. They also provide diagnostics and predict future failures.

The gas turbine is a highly responsive, high speed unit. In aircraft applications, a gas turbine can accelerate from idle to maximum take-off power in less than 60 seconds. In industrial gas turbines, the acceleration rate is limited by the mass of the driven equipment.

Without the proper control system, the compressor can go into surge in less than 50 milliseconds and the turbine can exceed safe temperatures in less than a quarter of a second. A power turbine can go into overspeed in less than two seconds.

Improvements in the properties of creep and rupture strength were steady from the late 1940s through the early 1970s. In 1950 390°F

(200°C) was achieved in operating temperature. This resulted from age-hardening and precipitation strengthening which utilized aluminum and titanium in the nickel matrix to increase strength. Since 1960, sophisticated cooling techniques have been used for turbine blades and nozzles. Since 1970, turbine inlet temperatures have increased to 500°F (260°C) and in some units as high as 2,640°F (1,450°C). The increase in turbine inlet temperature is possible with new air cooling techniques and the use of complex ceramic core bodies for hollow, cooled cast parts. The turbine blades and nozzles are formed by investment casting. A critical part is the solidification of the liquid metal after it is poured into the mold. Undesirable grain sizes, shapes, and transition areas can cause premature cracking of turbine parts. In the equiaxed process, uniformity of the grain structure occurs. Strength is improved if grain boundaries are aligned in the direction normal to the applied force.

This elongated or columnar grain formation in a preferred direction is called directional solidification. It was introduced by Pratt & Whitney Aircraft in 1965.

CONTROL

Control of the gas generator turbine and power-extraction turbine takes place by varying the gas generator speed, which is accomplished by varying the fuel flow. The following parameters may be monitored; fuel flow, compressor inlet and discharge pressures, shaft speed, compressor inlet temperatures, turbine inlet and exhaust temperatures (Figure 6-5). At a constant gas generator speed, as ambient temperature decreases, the turbine inlet temperature will decrease slightly and the gas horsepower will increase significantly. This increase in gas horsepower results from the increase in compressor pressure ratio and aerodynamic loading. This means the control must protect the gas turbine on cold days from overloading the compressor airfoils and overpressurizing the compressor cases. To get maximum power on hot days it is necessary to control the turbine inlet temperature to constant values, and allow the gas generator speed to vary.

The control senses ambient inlet temperature, compressor discharge pressure and gas generator speed. These are the three main variables that

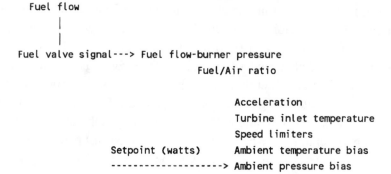

Figure 6-5. Gas turbine-generator control

affect the amount of power that the engine will produce. Sensing the ambient inlet temperature also helps to insure that the engine internal pressure are not exceeded, and sensing the turbine inlet temperature insures that maximum allowable turbine temperatures are not exceeded. Sensing the gas generator speed allows the control to accelerate through critical speed points. Gas turbines are typically flexible shaft machines and have a low critical speed.

The controls fall into several groups: hydromechanical (pneumatic or hydraulic), electrical (wired relay logic), and computer (programmable logic controller or microprocessor). Typical hydromechanical controls include cams, servos, speed (fly-ball) governors, sleeve and pilot valves, metering valves and temperature sensing bellows.

Electrical type controls include electrical amplifiers, relays, switches, solenoids, timers, tachometers, converters, and thermocouples. Computer controls may incorporate many electrical functions such as amplifiers, relays, switches, and timers. These functions are programmable.

This flexibility in modifying the program may be done by the user or operator in the field. Analog signals such as temperature, pressure, vibration, and speed are converted to digital signals before they are processed. The computer output signals to components such as the fuel valve, variable geometry actuator, bleed valve and anti-icing valve must be converted from digital to analog formats.

Until the late 1970s, control systems operated only in real time with no ability to store or retrieve data. Hydromechanical controls had to be calibrated frequently, weekly in some cases, and were subject to contamination and deterioration due to wear.

Multiple outputs such as fuel flow control and compressor bleed-air flow-control required independent, control loops. Coordinating the output of multiple loops, using cascade control, was a difficult task and often resulted in a compromise between accuracy and response time.

Many tasks had to be done manually. Station valves, prelude pumps and cooling water pumps were manually switched on before starting the gas generator. Protection devices were limited and the margin between temperature control setpoints and safe operating turbine temperatures had to be made large since hydromechanical controls cannot react quick enough to limit high turbine temperatures, or to shutdown the gas generator, before damage may occur.

In the early 1970s, electric controls consisted of a station control, a process control, and a turbine control. All control functions such as start, stop, load, unload, speed, and temperature were generated, biased, and computed electrically. Output amplifiers were used to drive servo valves, using high pressure hydraulics, to operate hydraulic actuators. These actuators may also contain position sensors to provide electronic feedback. The advent of programmable logic controllers and microprocessors in the late 1970s eliminated these independent control loops, and allowed multi-function control. Control system functions include sequencing, routine operations and protection.

Sequencing steps are used to start, load, unload, and stop the unit. The typical cycle used in a normal start are shown in Table 6-1.

When the start sequence is complete, the gas turbine will have reached self-sustaining speed and control is taken over by the routine operation controller. This controller maintains operation until it receives an input from the operator, or the process to load the unit. Before initiating this loading, control is turned over to the sequence controller to position the inlet and discharge valves and circuit breakers. In electric generator drives this is when the synchronizer is used to synchronize the unit to the electric grid. When these steps are completed, control is turned back over to the routine operation controller. At this time, the speed control governor, acceleration scheduler, temperature limit controller, and pressure limit controller become active. Table 6-2 lists gas turbine generators.

Table 6-1. Starting Cycle for Gas Turbine—10 to 30 Seconds

Starter on
Fuel on
Engine lights up - Exhaust gas temperature rise
Engine attains self-accelerating speed
Ignition off
Starter cuts out - Peak starting exhaust gas temperature rise
Engine attains idle RPM - Exhaust Gas Temperature drops to idle

Varying the fuel flow results in higher or lower combustion temperatures. As the fuel flow is increased, combustor heat and pressure increase and heat energy to the turbine is increased. Part of this increased energy is used by the compressor-turbine to increase speed which causes the compressor to increase airflow and pressure. The remaining heat energy is used by the power extraction turbine to produce more shaft horsepower. This cycle continues until the desired shaft horsepower or some parameter limit such as temperature or speed is reached.

To reduce shaft horsepower, the control reduces the fuel flow. The lower fuel flow reduces combustion heat and pressure. As the heat energy available to the compressor-turbine drops, the compressor-turbine slows down lowering the compressor speed as well as airflow and pressure. The downward movement continues until the desired shaft horsepower is reached.

If the fuel flow is increased too quickly, then excessive combustor heat is generated and the turbine inlet temperature may be exceeded or this increase in speed may drive the compressor into surge.

If the fuel flow is decreased too rapidly, then the compressor may not be able to reduce airflow and pressure fast enough. This can result in a flame-out or compressor surge, since speed decreases move the compressor operating point closer to the surge line. High turbine inlet temperature will shorten the life of the turbine blades and nozzles, and compressor surge can severely damage the compressor blades and stators

Table 6-2. Typical Gas Turbine Generators—Combined Cycle

Company/ Model	Frequency	ISO Rating Output (MW)	Efficiency	Gas Turbine Output (MW)	Steam Turbine Output (MW)
ABB Power					
KA26-1	50	366	58.5	232.6	133.4
KA35-2	50/60	45.8	43.3	33.2	12.6
Ansalado Energia					
164.3	50/60	90.4	51.6	60.6	32
294.3A	50	706	57	466	249
GHH Borsig Turbomaschnen GMBH					
FT8	50/60	32.8	48.6	25.7	7.5
FT8 Twin	50/60	68.9	59.4	51.6	18.5
Ebara Corp.					
FT8 Power	50/60	32.2	48.7	24.7	7.58
FT8 Twin	50/60	65.3	49.2	49.6	15.6
Fiatavio Per L'Energia					
CC130	50	127.2	48.85	50.0	38.6
CC30	60	32.8	52.56	10.2	22.5
Gec Alsthom					
VEGA 105	50/60	38.7	41.9	25.9	13.3
VEGA 209FA	50	705.4	55.5	448.8	265.7
GE Power Systems					
S106B	50/60	59.2	48.6	37.5	22.6
S260	60	105.0	52.4	77	30
Hitachi Limited	50/60	69.1	47.6	44.8	24.2
	50/60	471.7	47.6	227.8	118.7

(*Continued*)

Table 6-2. (Continued)

Japan Gas Turbine KK					
	50/60	163.8	52.1	105.5	58.3
	50/60	924.4	57.9	465.2	259.2
Kawasaki Heavy Industries Limited					
KA13E2-1	50	238.4	52.0	158.5	79.9
M7A-01	50/60	16.3	41.9	11.0	52.9
Mitsubishi Heavy Industries Limited					
MPCP1(701)	50	194.5	50.5	128.7	65.8
MPCP2(501G)	60	688.9	58.2	455.4	233.5
Mitsui Engineering & Shipbuilding Company					
SB30	50/60	770	38.1	506	264.0
SB120	50/60	32.8	42.4	22.34	10.4
Rolls Royce					
	50/60	36.6	42.2	26.4	11.0
	50/60	655.3	51.9	449.8	221.9
Siemens AG Power Generation Group					
GUD 1.64.3	50/60	90.0	51.5	61	31
GUD 2.94.3A	50/60	705	57	466	249
Solar Turbines					
IPS30	50/60	29	42	20.7	9.2
IPS60	50/60	70.3	42	51.7	22.7
Thomassen International BV					
STEG 106 B-DP	50/60	61.7	50.1	39.2	22.5
STEG-LM160	50/60	52.1	51.9	38.5	13.6
Turbo Power & Marine					
FTB Power Pac	50/60	32.8	49.3	25.0	839.5
FTB Twin Pac	50/60	67.0	50.2	50.4	17.7
Turbotechnical					
CC-201	50/60	28.3	44.5	19.0	10.0
CC-260	50	103.0	51.4	75.2	30.0
Westinghouse Electric Corp.					
RB211	50/60	37.3	50.59	26.2	11.7
701F	50	713.3	55.22	478.0	243.9

and possibly the rest of the gas turbine. Flame-out creates thermal stresses that become critical with each shutdown and re-start.

COMPRESSOR SURGE

The control must also guard against surge during rapid power changes, start-up, and periods of operation when the compressor inlet temperature is low or drops rapidly. The gas turbine is more susceptible to surge at low compressor inlet temperatures.

Normally, changes in ambient temperature are slow compared to the response time of the gas turbine control system. The temperature range from 28°F (-2.0°C) to 42°F (6.0°C) with high humidity is a major concern. Operation in this range can result in ice formation in the plenum upstream of the compressor. Anti-icing schemes increase the sensible heat by introducing hot air into the inlet. Anti-icing is another control function that must address the effect temperature changes can have on compressor surge.

An acceleration schedule loads the unit as quickly as possible. As the load goes from the idle-no-load position to the full-load position, the fuel valve is opened and as the load approaches the setpoint, the speed governor begins to override the acceleration schedule output until the fuel valve reaches its final running position. During this time the temperature limit controller and the pressure limit controller monitor temperatures and pressures so that the preset levels are not exceeded.

The temperature limit controller for turbine inlet temperature uses the average of several thermocouples taking temperature measurements in the same plane. When the temperature or pressure reaches its setpoint, the limit controller will override the governor controller and maintain operation at a constant temperature or pressure. The control allows the operating point to move along a set of points that define the operating line for the load conditions.

A protection controller continuously checks the speed, temperature, and vibration for levels that may be harmful to the operation of the unit. Usually two levels are set for each parameter, an alarm level and a shutdown level. When the alarm level is reached, the system provides a warning that there is a problem. If the transition from alarm to shutdown condition takes place too rapidly and operational response is not possible, the unit is automatically shutdown. Overspeed is one parameter that does not include an alarm signal.

Turbine inlet temperature is the most frequently activated limiting factor. One level is set for base load operation and a higher level is set for peaking operation.

AUXILIARY EQUIPMENT

Auxiliary turbine equipment includes the starting system, ignition system, lubrication system, air inlet cooling system, water or steam injection system for NO_x control or power augmentation and the ammonia injection system for NO_x control.

These systems may be direct driven and connected directly to the shaft of the gas turbine. In most cases, one of the lubrication pumps is direct connected. Indirect drives use electric, steam, or hydraulic motors for power. Indirect drives allow redundant systems, increasing the reliability. Electric systems are powered by a directly driven electric generator.

Starting systems may drive the gas generator directly or through a gearbox. Starters may be diesel or gas engine, steam or gas turbine, electric, hydraulic, or pneumatic air or gas. The starter must rotate the gas generator until it reaches its self-sustaining speed and drive the gas generator compressor to purge the gas generator and the exhaust duct of volatile gases before initiating the ignition cycle. The starting sequence engages the starter, purges the inlet and exhaust ducts, energizes the ignitors and switches the fuel on.

The starting system must accelerate the gas generator from rest to a speed just beyond the self-sustaining speed of the gas generator. The starter must develop enough torque to overcome the drag torque of the gas generator's compressor and turbine, the attached load including accessories and bearing resistance. The single shaft gas turbines with directly attached loads used in electric generators represent the highest starting torque.

Another function of the starting systems is to rotate the gas generator, after shutdown, to begin cool down. The purge and cooldown functions have resulted in the use of two-speed starters. The lower speed is used for purge and cooling and the higher speed is used to start the unit.

Gas generators are started by rotating the compressor. The starter may be directly connected to the compressor shaft or indirectly connected with an accessory gearbox or impingement air may be directed into the compressor-turbine. Starters for gas generators include alternating current and direct current motors, pneumatic motors, hydraulic motors, diesel motors and small gas turbines.

If alternating current (AC) power is available, three-phase induction type motors are preferred as drivers. The induction motors is directly connected to the compressor shaft or the starter pad of the accessory gearbox. Once the gas generator reaches self-sustaining speed, the motor is de-energized and usually disengaged through a clutch mechanism.

If AC power is not available (black start applications), direct current (DC) motors are used. The source of power is a battery bank. One approach is to convert the DC motor electrically into a electric generator to charge the battery system. This is useful where the battery packs are also used to provide power for other systems. Battery-powered DC motor starters are mostly used in small, self-contained gas turbines under 500 brake horsepower (BHP).

Electric motors require explosion proof housings and connectors and must be rated for the area classification in which they are installed. Typically this is Class 1, Division 2, Group D.

Pneumatic starter motors may be the impulse-turbine or vane pump type. These motors use air or gas as the driving force, and are coupled to the turbine accessory drive gear with an overriding clutch. The overriding clutch mechanism disengages when the drive torque reverses and the gas turbine self-accelerates faster than the starter. Then, the air supply is shut-off. Air or gas must be available at approximately 100 psig and in sufficient quantity to sustain starter operation until the gas generator exceed self-sustaining speeds.

If a continuous source of air or gas is not available, banks of high and low pressure receivers and a small positive displacement compressor can provide air for a limited number of start attempts. The starting system should be capable of three successive start attempts before the air supply system must be recharged. In gas pipeline applications, the pneumatic starter may use pipeline gas as the source of power.

Hydraulic pumps can provide the power to drive hydraulic motors or hydraulic impulse turbine, Pelton Wheel starters. Hydraulic systems are often used with aircraft derivative gas turbines.

Large heavy frame, 25,000 SHP and above, gas turbines require high torque starting systems. Most of these units are single shaft machines and the starting torque must be sufficient to overcome the mass of the gas turbine and the driven load. Diesel motors are preferred for these large gas turbines. Since diesel motors cannot operate at gas turbine speeds, a speed increaser gearbox is used to boost the diesel motor starter speed to gas turbine speeds. Diesel starters are usually connected to the compressor shaft. Besides the speed increaser gearbox, a clutch mechanism is needed to insure that the starter can be disengaged from the gas turbine. Diesel motors can run on the same fuel as the gas turbine, eliminating the need for separate fuel supplies.

Small gas turbines may be used to provide the power to drive either pneumatic or hydraulic starters. In aircraft, a combustion starter, which is essentially a small gas turbine, is used to start the gas turbine in remote locations. They are not used in industrial applications.

Impingement starting uses jets of compressed air piped to the compressor turbine to rotate the gas generator. The pneumatic power source for impingement starting is similar to air starters.

Ignition is only required during start-up. Once the unit has accelerated to self-sustaining speed the ignition system can be de-energized. The ignition system is not energized until the gas generator reaches cranking speed and remains at this speed long enough to purge volatile gases from the engine and exhaust duct.

When the igniters are energized, fuel can then be admitted into the combustor. These two functions are often done simultaneously and called pressurization.

Two igniters are usually used, one on each side of the engine. During the start cycle each igniter discharges about 2 times per second and provides an energy pulse of 4 to 30 joules. A joule is the unit of work or energy transferred in one second by an electric current of one ampere in a resistance of one ohm. One joule/second equals one watt.

Once the gas generator starts the igniter is no longer needed and any further exposure to the hot gases of combustion shortens its life. Some igniter are spring loaded and retract out of the gas path as the combustion pressure increases.

The potential at the spark plug is about 25,000 volts. The ignition harness to each igniter plug is shielded and the ignition exciter is hermetically sealed.

Ignition systems include inductive and capacitive AC and DC, with high and low tension systems. The capacitive systems generate the hottest spark. Since the energy stored in the capacitor is proportional to the square of the voltage, it is more economical to use a high voltage to charge the capacitor. A radio frequency interference filter is used to prevent ignition energy from affecting local radio signals.

An AC transformer, or transistorized chopper circuit transformer, boosts the voltage to about 2,000 volts in the low tension system (Figure 6-6). A rectifier allows the flow of current into the storage capacitor but prevents most of the return flow. This voltage is boosted up to charge a smaller high tension capacitor.

The low voltage charge in the storage capacitor is not enough to jump the gap across the spark plug electrodes. The initial path is provided by a higher voltage discharge from the high tension capacitor (Figure 6-7). It discharges first to bridge the gap across the electrodes of the spark plug and reduces the resistance for the low tension discharge. The low tension capacitor then discharges providing a long, hot spark.

The inductive ignition system uses the rapid variation in magnetic flux in an inductive coil to generate enough energy for the spark (Figure

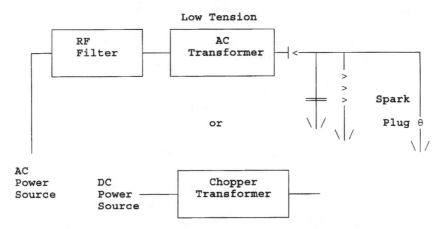

Figure 6-6. Low voltage ignition system

6-8). This system produces a high voltage spark, but the energy is relatively low and is only suitable for easily ignitable fuels.

Ignitor plugs use an annular-gap or constrained-gap. The annular-gap plug projects slightly into the combustor, while the constrained-gap plug is positioned in the plane of the combustor liner and operates in a cooler environment.

Lubrication systems provide lubrication between the rotating and stationary bearing surfaces and remove excessive heat from those surfaces. The bearings may be hydrodynamic or anti-friction types. The lubrication in a hydrodynamic bearing converts sliding friction into fluid friction. Anti-friction bearings work on rolling friction. The shaft load is

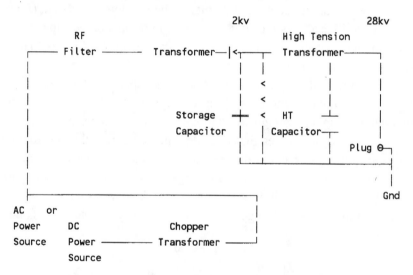

Figure 6-7. High energy capacitor ignition system

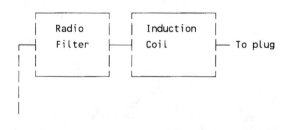

Figure 6-8. DC inductive-type ignition system

supported by the rolling elements and races in a metal-to-metal contact.

Most heavy frame gas turbines use hydrodynamic bearings with mineral oil while the aircraft derivative gas turbines use anti-friction bearings with a synthetic oil. Mineral oils are distilled from petroleum crude oils and are generally less expensive than synthetic lubricants. Synthetic lubricants do not occur naturally but are made by reacting organic chemicals, such as alcohol or ethylene with other elements. Synthetic lubricants are used in high temperature applications of less than 350°F (175°C) or where fire-resistant qualities are required.

The bearings in gas turbines are lubricated by a pressure circulating system. This consists of a reservoir, pump, regulator, filter, and cooler. Oil in the reservoir is pumped under pressure through a filter and oil cooler to the bearings and then returned to the reservoir for re-use. In cold climates a heater in the reservoir warms the oil prior to start-up.

The reservoir also serves as a deaerator. As the lubrication oil moves through the bearings, it can entrap air in the oil. This results in oil foaming. The foam must be removed before the oil is returned to the pump or the air bubbles can result in pump cavitation. To deaerate the oil the reservoir surface area must be large enough so screens and baffles may also be built into the reservoir.

Filters are used to remove bearing wear particles from the oil. While 5 or 10 micron filters can be used for running conditions, 1 to 3 micron filters are used for break-in periods or after overhauls. Redundant oil filters are used along with a three-way-transfer valve. If the primary filter clogs, the transfer valve is switched over to the clean filter.

Wear particles from the pump and gas turbine bearings will accumulate in the filter element along with temperature related oil degradation and oil additives. This can create a sludge that accumulates in the filter. As the filter clogs, the differential pressure across the filter increases. Instrumentation includes a pressure differential gauge for local readout, and a differential pressure transducer for remote readout and alarm. The differential pressure alarm setting is typically 5 psig.

Regulators are used to maintain a constant pressure level in the lube system. These regulators allow the operation of a secondary pump for preventive maintenance.

Lube oil coolers remove heat from the oil before it is re-introduced into the gas turbine. The amount of cooling required depends on the friction heat generated in the bearings, heat transfer from the gas turbine to the oil by convection and conduction and heat transfer from the hot gas path through seal leakage. The oil is cooled to 120°F-140°F (50°C-60°C). To maximize the heat transfer, fins are installed on the outside of tubes and turbulators are placed inside each cooling tube. The turbulators help to transfer heat from the hot oil to the inner wall of the cooling tubes and the fins help dissipate this heat. Cooling media may be either air or a water/glycol mix. Air/oil coolers are used in desert regions, while tube and shell coolers are found in the Arctic regions and most coastal regions.

Air/oil coolers use ambient air as the cooling media. Cooling fans are usually electric motor driven, often with two speed motors. This allows high and low cooling flows. To closely match the cooling flow to the required heat load, changeable pitch fan blades may be used. As the heat load changes, the blades can be adjusted to meet the new heat flow requirements. Air/oil coolers may also include top louvers to protect the cooling coils from hail. These louvers are not effective for temperature control.

ENERGY EFFICIENT COGENERATION SYSTEMS

Cogeneration can offer an effective way to conserve energy while reducing energy costs. However, a cogeneration plant is capital intensive and the matching process is critical. Development is both plant-specific and site-specific.

An industry may use fossil fuel to produce heat and will buy electricity from the local utility. The fuel will be burned to produce hot gases for direct applications, such as drying or indirect applications such as steam generators or boilers and combustion engines doing mechanical work, such as gas turbines.

In all of these conversion processes, only a part of the original heat content of the fuel is utilized. The remainder is rejected. Unless this reject heat is put to good use, it can contribute only to global entropy increase.

Combustion gases are carriers of energy and supply the conversion equipment with available heat. This is the heat content which can be converted to work by reducing the temperature of the carrier medium.

STEAM TURBINES

One of the most common methods of cogenerating electricity involves using excess steam from industrial processes (Figure 6-9). Steam turbines are popular drives for rotary equipment and are second only to electric motors. Some steam turbines can only be used for constant speed applications because of the characteristics of their steam supply valves. Their governors are not suitable for throttling.

Steam turbines are energy conversion machines. They extract energy from steam and convert it into shaft energy. Sizes range from a few kW to over 1000-MW. Rotational speeds vary from 1,800 to 12,000 rpm. Gas turbines run at speeds up to 18,000 rpm when driving a compressor or generator.

The amount of energy that can be extracted from the steam depends on the enthalpy drop across the machine. The enthalpy of the steam in turn depends on the temperature and pressure. In order to operate at maximum efficiency, the steam should hit the turbine blades at sonic velocities. The efficiency of steam turbines increases with size, the superheat temperature of the steam and the degree of vacuum at the turbine exhaust.

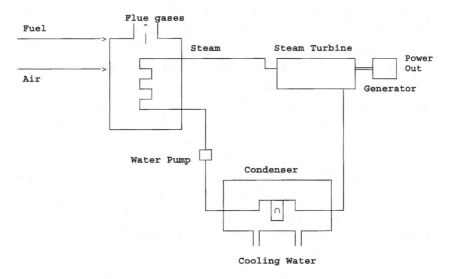

Figure 6-9. Conventional steam turbine

Steam turbines may be condensing or noncondensing. In condensing turbines the steam is condensed at the exhaust of the turbine. The exhaust pressure is subatmospheric. In systems with these turbines, the steam used for process must be extracted before the turbine's exhaust. Extraction valves are used to supply the process steam.

Noncondensing turbines produce power by operating as a pressure reducer. Noncondensing units are also called backpressure units where the exhaust pressure is greater than atmospheric. The turbine exhaust becomes low pressure process steam. Noncondensing turbines are less expensive to buy and operate, since the energy is extracted from the steam while it has a higher enthalpy and smaller volume per unit of energy. This reduces the size of the turbine.

Turbines may have a steam flow that is described as axial or radial and a small number even have tangential flow. A unit may be single-stage or multistage. The number of parallel exhaust stages can also be used to describe turbines such as single or double flow.

Generating electricity in these systems depends on steam flow which may not increase or decrease to meet the demanded load. In a condensing turbine system, the boiler pressure can be increased to provide enough electricity to meet the demand. Condensing turbines are not usually available smaller than 5,000 kW, since they are less efficient than noncondensing units. The heat of the steam is lost once it leaves the turbine. They are also more expensive.

The cost range for a 1-MW steam turbine is about $130-330/kW. For 25-MW turbines this drops to less than $100/kW. These costs are for noncondensing systems without peripheral items, such as piping and electrical switchgear. They do include turbines, generators, baseplates and lube systems. Maintenance for these systems usually runs about 0.3 to 0.4 cents per kilowatt hour.

Electricity generated by steam turbine cogeneration systems can be about 1/3 less expensive than electricity from a utility. Steam turbine cogeneration systems are generally not cost efficient below the 1-MW range. Systems above 20-MW have been popular for many years, because of their cost savings.

BIOFUELS

Biofuels come from biomass products which may be energy crops, forestry and crop residues and even refuse. One feature of biofuels is that 3/4 or more of their energy is in the volatile matter or vapors unlike coal, where the fraction is usually less than half. It is important that the furnace or boiler ensure that these vapors burn and are not lost.

For complete combustion, air must reach all the char, which is achieved by burning the fuel in small particles. This finely-divided fuel means finely-divided ash particulates which must be removed from the flue gases.

The air flow should be controlled. Too little oxygen means incomplete combustion and leads to the production of carbon monoxide. Too much air is wasteful since it carries away heat in the flue gases. Modern systems for burning biofuels include large boilers with megawatt outputs of heat.

Direct combustion is one way to extract the energy contained in household refuse, but its moisture content tends to be high at 20% or more and its energy density is low. A cubic meter contains less than 1/30 of the energy of the same volume of coal.

Refuse-derived fuel (RDF) refers to a range of products resulting from the separation of unwanted components, shredding, drying and treating of raw material to improve its combustion properties. Relatively simple processing can involve separation of large items, magnetic extraction of ferrous metals and rough shredding. The most fully processed product is known as densified refuse-derived fuel (d-RDF). It is the result of separating out the combustible part which is then pulverized, compressed and dried to produce solid fuel pellets with about 60% of the energy density of coal.

Anaerobic digestion, like pyrolysis, occurs in the absence of air. But, the decomposition is caused by bacterial action rather than high temperatures. This process takes place in almost any biological material, but it is favored by warm, wet and airless conditions. It occurs naturally in decaying vegetation in ponds, producing the marsh gas that can catch fire.

Anaerobic digestion also occurs in the biogas that is generated in sewage or manure as well as the landfill gas produced by refuse. The resulting gas is a mixture consisting mainly of methane and carbon dioxide.

Bacteria breaks down the organic material into sugars and then into acids which are decomposed to produce the gas, leaving an inert residue whose composition depends on the feedstock.

The manure or sewage feedstock for biogas is fed into a digester in the form of a slurry with up to 95% water. Digesters range in size from a small unit of about 200 gallons to ten times this for a typical farm plant and to as much as 2000 cubic meters for a large commercial installation. The input may be continuous or batch. Digestion may continue for about 10 days to a few weeks. The bacterial action generates heat but in cold climates additional heat is normally required to maintain a process temperature of about 35°C.

A digester can produce 400 cubic meters of biogas with a methane content of 50% to 75% for each dry ton of input. This is about 2/3 of the fuel energy of the original fuelstock. The effluent which remains when digestion is complete also has value as a fertilizer.

A large proportion of municipal solid wastes (MSW), is biological material. Its disposal in deep landfills furnishes suitable conditions for anaerobic digestion. The produced methane was first recognized as a potential hazard and this led to systems for burning it off. In the 1970s some use was made of this product.

The waste matter is miscellaneous in a landfill compared to a digester and the conditions not as warm or wet, so the process is much slower, taking place over years instead of weeks. The product, called landfill gas (LFG), is a mixture consisting mainly of CH_4 and CO_2.

A typical site may produce up to 300 cubic meters of gas per ton of wastes with about 55% by volume of methane. In a developed site, the area is covered with a layer of clay or similar material after it is filled, producing an environment to encourage anaerobic digestion. The gas is collected by pipes buried at depths up to 20 meters in the refuse.

In a large landfill there can be several miles of pipes with as much as 1000 cubic meters an hour of gas being pumped out. The gas from landfill sites can be used for power generation. Some plants use large internal combustion engines, standard marine engines, driving 500-kW generators but gas turbines could give better efficiencies.

Methanol can be produced from biomass by chemical processes. Fermentation is an anaerobic biological process where sugars are converted to alcohol by micro-organisms, usually yeast. The resulting alcohol is ethanol. It can be used in internal combustion engines, either directly in modified engines or as a gasoline extender in gasohol. This is gasoline containing up to 20% ethanol.

One source of ethanol is sugarcane or the molasses remaining after the juice has been extracted. Other plants such as potatoes, corn and other grains require processing to convert the starch to sugar. This is done by enzymes.

The fuel gas from biomass gasifiers could be used to operate gas turbines for local power generation. A gas-turbine power station is similar to a steam plant except that instead of using heat from the burning fuel to produce steam to drive the turbine, it is driven directly by the hot combustion gases. Increasing the temperature in this way improves the thermodynamic efficiency, but in order not to corrode or foul the turbine blades the gases must be very clean which is why many gas-turbine plants use natural gas.

In systems where the wastes from processing vegetable material provide the input, process heat (hot steam) as well as electric power may be the output. For these systems the steam-injected gas turbine (STIG) is well-suited.

This type of turbine is driven by both combustion gases and high-pressure steam and can operate with flexibility by responding to varying demands for heat. The total system is a biomass integrated gasifier/steam-injected gas turbine.

Advances in gas turbine technology have been driven by the aircraft industry, and the development of reliable gasification systems for clean coal combustion, have resulted in these flexible biomass electricity systems of 20-100-MW.

Steam-injected gas turbines appeared in the U.S. in the 1980s (Figure 6-10). Originally, they were fired by natural gas and then adapted for the gas from coal gasification. An advanced form of this technology is intercooled steam injection. Conventional sugar mills can generate 15-25-kWh of electricity per ton of cane crushed. Electricity generation with steam turbines can produce up to 100-kWh per ton of cane. Steam injection technology can boost this to 280-kWh a ton.

One 37-MW combined-cycle gas turbine system based on conventional combined-cycle gas turbine technology and an air-blown Ahistrom gasifier should achieve a 42% overall conversion efficiency.

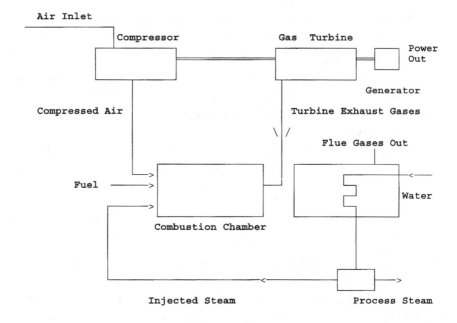

Figure 6-10. Steam-injected gas turbine

GASIFICATION

Gasification includes a range of processes where a solid fuel reacts with hot steam and air or oxygen to produce a gaseous fuel (Figure 6-11). There are several types of gasifiers, with operating temperatures ranging from a few hundred to over a 1000°C, and pressures from near atmospheric to as much as 30 atmospheres.

The resulting gas is a mixture of carbon monoxide, hydrogen, methane, carbon dioxide and nitrogen. The proportions depend on the processing conditions and if air or oxygen is used.

Gasification is not new. Town gas is the product of coal gasification and was widely used for many years before it was displaced by natural gas. Many vehicles towing wood gasifiers as their fuel supply were used in World War II.

The growth of interest in biomass gasification in recent years may result in a fuel which is much cleaner than the original biomass, since undesirable chemical pollutants can be removed during the processing, together with the inert matter which produces particulates (smoke) when

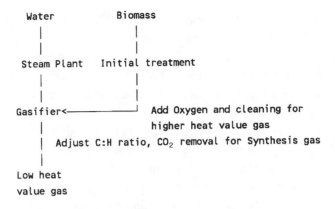

Figure 6-11. Gasification process

the fuel is burned. The resulting gas is a much more versatile fuel. Besides direct burning for heating, the gas can also be used in internal combustion engines or gas turbines. Gasification under the proper conditions can also produce synthesis gas, a mixture of carbon monoxide and hydrogen which can be used to synthesize almost any hydrocarbon.

The simplest process results in gas containing up to 50% by volume of nitrogen and CO_2. Since these have no fuel value, its energy is only a few megajoules per cubic meter or about a tenth of that of methane, but it is a clean fuel.

Pyrolysis is the simplest and probably the oldest method of processing one fuel in order to produce a better one. Conventional pyrolysis involves heating the original material in the near-absence of air, typically at 300-500°C, until the volatile matter has been driven off. The residue is commonly known as charcoal, a fuel which has about twice the energy density of the original and burns at a much higher temperature.

In much of the world today, charcoal is produced by pyrolysis of wood. Depending on the moisture content and the efficiency of the process, 4-10 tons of wood are required to produce one ton of charcoal, and if no attempt is made to collect the volatile matter, the charcoal is obtained at the cost of perhaps two-thirds of the original energy content.

In more sophisticated pyrolysis techniques, the volatiles are collected, and careful control of the temperature controls their composition. The liquid product can be used as fuel oil, but is contaminated with acid and must be treated before use. Fast pyrolysis of plant material, such as wood or nutshells, at temperatures of 800-900°C converts about 10% of the material to solid charcoal and about 60% into a gas of hydrogen and carbon monoxide. Lower temperatures allow fewer potential pollutants are emitted than in full combustion. Small-scale pyrolysis plants have been used for treating wastes from the plastics industry as well as disposing of used tires.

A gasifier which uses oxygen instead of air produces a gas that consists mainly of H_2CO and CO_2. Removing the CO_2 leaves a mixture called synthesis gas, from which almost any hydrocarbon compound may be synthesized including methane with an energy density of 23-GJ per ton.

Methanol can be used as a substitute for gasoline. The technology has been developed for use with coal as the feedstock by coal-rich countries at times when their oil supplies were threatened.

COMBINED CYCLE TECHNOLOGY

The conversion of coal to a gaseous fuel can be done using coal gasification. Then the application of combined cycle technology can be used for the production of power using the coal gas. This is called integrated gasification-combined-cycle (IGCC).

Most recent domestic powerplants are based on either simple or combined cycle natural gas fueled combustion turbines. Coal-fired powerplants are still a major source of power worldwide. The emissions produced by coal combustion have led to environmental concerns. This has led to the installation of flue gas scrubbers in conventional coal-fired powerplants.

IGCC is a power source concept which is not only more acceptable from an environmental standpoint, but also provides higher efficiency. Applying combined cycle technology to power production from coal requires conversion of coal to a gaseous fuel via a coal gasification process. Coal gasification processes involve partial combustion of the coal to provide energy for further conversion of the coal into a gaseous fuel. This fuel's primary components are carbon monoxide, hydrogen and nitrogen.

In a coal gasification system, partial combustion takes place in the gasifier and is completed in the gas turbine combustors. Particulates and sulfur are removed by cyclones and a HGCU unit between these stages. The final fuel arriving at the machine, although of low heating value, requires relatively few changes to the gas turbine.

An IGCC system is used at the Sierra Pacific Power Company's Pinon Pine Powerplant. It includes the integration of an advanced technology gas turbine in an air-blown gasifier. This 800-ton-per-day air-blown integrated gasification combined cycle plant is at the Sierra Pacific Power Company's Tracy Station near Reno, Nevada. A General Electric MS6001FA gas turbine/air-cooled generator and steam turbine/TEWAC generator is used with a combined cycle output of about 100-MW.

This was the first air-blown IGCC powerplant to incorporate an F-technology gas turbine generator. This type of combustion turbine has a high combined cycle efficiency which enhances the economics of an IGCC application.

A simplified IGCC system incorporates air-blown gasification with hot gas cleanup. This eliminates the oxygen plant and minimizes the need for gas cooling and wastewater processing equipment. The existing Tracy Station is a 400-MW, gas/oil-fired power generation facility about 20 miles east of Reno at an elevation of 4280 feet above sea level.

A Kellogg-Rust-Westinghouse (KRW) ash-agglomerating fluidized-bed gasifier operates in the air-blown mode and is coupled with hot gas cleanup (HGCU) will provide a low heating value fuel gas to power the combustion turbine. High temperature exhaust from the combustion turbine supplies the energy required to generate steam in a heat recovery steam generator (HRSG) for use in a steam turbine. Both the combustion turbine and the steam turbine drive generators supply approximately 100-MW of power to the electric grid.

The pressurized (20 bars) KRW gasifier has in-bed desulfurization, external regenerable sulfur removal and fine particulate filters. Advanced KRW gasification technology produces a low-Btu gas with a heating value of approximately 130 Btu/SCF. This is used as fuel in the combined-cycle powerplant, and includes hot gas removal of particulates and sulfur compounds from the fuel gas. Desulfurization is accomplished by a combination of limestone fed to the gasifier and treatment of the gas in desulfurization vessels using a zinc-based sulfur sorbent such as Z-Sorb.

Particulates are removed by a pair of high-efficiency cyclones and a barrier filter. These operations are carried out at the elevated temperature of approximately 1000°F (538°C) to eliminate thermodynamic inefficiency and the costs of cooling and cleaning the gas at low temperature, which is done in most IGCC systems. Since the water vapor is not condensed in the hot gas cleanup process, water effluents are reduced and contain only feed water treatment effluent and boiler and cooling tower blow down.

Sub-bituminous coal is received at the plant from a unit train consisting of approximately 84 railcars of between 100- and 110-ton capacity, arriving approximately once a week. Coal is received at an enclosed unloading station and transferred to a coal storage dome. The unloading station has two receiving hoppers, each equipped with a vibrating-type unloading feeder that feeds the raw coal conveyor systems.

The steam turbine is a straight condensing unit with uncontrolled extraction and uncontrolled admission. The steam conditions are 950 psia (65.5 Bars), 248°F (510°C) inlet, with exhaust at 0.98 psia (66m Bars). The uncontrolled extraction pressure is 485 psia (33.4 Bars), and the uncontrolled admission is 54 psia (3.7 Bars). The turbine has 26 inch (0.66m) last stage buckets in an axial exhaust configuration. The turbine is baseplate mounted, with a combined lubrication and control system mounted on a separate console. The control is a GE Mark V Simplex system. The generator is a baseplate mounted GE TEWAC unit rated at 59000 kVA at 0.85 power factor.

The gas turbine is a General Electric MS600FA. It is aerodynamically scaled from the larger MS7001FA machine. Key characteristics of the FA cycle are the 235°F (1288°C) firing temperature allowing high combined cycle efficiencies and the ability of the combustion system to burn a wide spectrum of fuels.

The MS6001FA has a compressor or cold-end drive flange, turning an open-ventilated GE 7A6 generator (82000 kVA @ 0.85PF) through a reduction load gear. The exhaust is an axial design directing the gas flow into the HRSG.

The MS6001FA uses an axial flow compressor with 18 stages, the first two stages are designed to operate in transonic flow. The first stator stage is variable. Cooling, sealing and starting bleed requirements are handled by 9th and 13th stage compressor extraction ports.

The rotor is supported on two tiltpad bearings. It is made up of two subassemblies, the compressor rotor and turbine rotor, which are bolted together. The 16-bladed disks in a through-bolted design provide stiffness and torque carrying capability. The forward bearing is carried on a stub shaft at the front of the compressor.

The three-stage turbine rotor is a rigid structure with wheels separated by spacers with an aft bearing stub shaft. Cooling is provided to wheel spaces, all nozzle stages and bucket stages one and two.

The main auxiliaries are motor driven and arranged in two modules. An accessory module contains lubrication oil, hydraulic oil, atomizing air, natural gas/doped propane skids and bleed control valves. The second module would normally house liquid fuel delivery equipment but in this application, holds the syngas fuel controls. A fire resistant hydraulic system is used for the large, high temperature syngas valves.

The fuel system is designed for operation on syngas, natural gas or doped propane. Only natural gas or doped propane can be used for start up with either fuel delivered through the same fuel system. Once the start is initiated it is not possible to transfer between natural gas and doped propane until the unit is on-line.

A full range of operation from full-speed/no-load to full-speed/full-load is possible on all fuels. Transfers, to or from syngas, can take place down to 30% of base load. This maintains a minimum pressure ratio across the fuel nozzle preventing the transmission of combustor dynamics or cross-talk between combustors. Cofiring is also possible with the same 30% minimum limit apply to each fuel. A low transfer load also minimizes the amount of syngas flaring required.

The gasifier has a fixed maximum rate of syngas production. At low ambient temperatures, fuel spiking is needed to follow gas turbine output.

The MS6001FA gas turbine can accept a limited amount of natural gas mixed with the syngas, to boost the Btu content. In this application the maximum expected is about 8% and an orifice in the natural gas mixing line will limit supply to 15% in the event of a valve malfunction. This protects the gas turbine from surges in the combustor. On hot days power augmentation can be achieved with steam injection or evaporative coolers.

The natural gas/doped propane fuel system, uses a modulating stop ratio valve to provide a constant pressure to a critical flow control valve. The position of the valve is then modulated by the speed control to supply the desired fuel flow to the combustors. When not in use, the natural gas/ doped propane system is purged with compressor discharge air.

Syngas is supplied from the gasifier at a temperature and pressure of approximately 1000°F (538°C) and 240 psia (16.5 Bars). The values depend on the operating conditions.

The specific heating value of the syngas can vary, so the delivery system is sized to accommodate the resulting volumetric flow changes. The low Btu content and elevated temperature of the syngas mean that large volumes of gas are needed. Flow from the gasifier takes place in a 16-inch (41cm) diameter pipe with a 12-inch (30cm) diameter manifold supplying syngas to the six combustors via 6-inch (15cm) diameter flexible connections. In contrast, the corresponding figures for natural gas are 4 (10), 3 (7.6cm) and 1.25 inches (3.1cm).

Syngas at 1000°F (538°C) burns spontaneously when in contact with air. The fuel system is designed to eliminate syngas-to-air interfaces. Both steam and nitrogen purging and blocking systems may be used.

Nitrogen has been used as a purge medium in other IGCC applications. While operating on syngas, nitrogen would be supplied at a pressure higher than the syngas and would use a block and bleed strategy to prevent contact with the purge air supply. Operating on natural gas, the syngas would be blocked by an additional stop valve and a stop ratio valve. The nitrogen purge valves open to flush syngas from the entire system downstream of the stop valve, and then close. Compressor discharge air is then used to purge the line from the control valve to the combustor while nitrogen provides a block and bleed function between the control valves. Sequencing of the valves maintains separation between syngas and air.

The turbine uses a reverse flow can-annular combustion arrangement typical of GE heavy duty gas turbines. There are six combustors, equally spaced and angled forward from the compressor discharge case. Compressor discharge air flow over and through holes in the impingement sleeve provide cooling to the transition piece. The remainder of the flow passes through holes in the flow sleeve. The air then flows along the annulus formed by the case and the liner.

The combustor uses conventional film cooling. Fuels and diluents are introduced through nozzles in the end cover. The combustor is the diffusion type.

A dual gas end cover is used with large syngas nozzles and natural gas/doped propane nozzles for the alternate fuel. Steam injection nozzles are used for power augmentation at high ambient temperatures and NO_x control.

The combustion cases have provisions for flame detectors and cross-fire tubes, along with extraction ports to supply compressor discharge air to the gasifier.

CONTROL SYSTEM

The plant is controlled by a Distributed Control System (DCS). The turbine control system interfaces with the DCS. The natural gas/doped propane fuel system provides a signal which indicates which fuel is in use. Logic in the control system makes the necessary specific gravity adjustments for fuel flow measurement.

The syngas control system provides an integrated control system for the syngas boost compressor and the gas turbine to minimize fuel supply delivery pressure losses. The boost compressor raises the compressor extraction pressure prior to the gasifier. After the gasifier, fuel is metered to the gas turbine through the stop ratio and gas control valves. Pressure drops through these valves are minimized by controlling them to the full open position during normal control operation. In this way the system operates at the minimum pressure drop and controls the gasifier flow via the boost compressor by controlling the gas turbine output. When the fuel flow is not meeting demand, a signal is sent to the gasification control panel, which translates this input to the necessary response from the boost compressor/gasifier. Syngas output is increased or decreased as demanded while the stop ratio and gas control valves remain wide open.

The control handles fuel transfers, co-firing, purge/extraction valve sequencing, steam injection, surge protection along with the standard gas turbine control functions.

WASTE HEAT GAS TURBINES

Gas turbines are used for cogeneration in refineries. The fuels used to generate steam, in most cases, are generated internally in the refinery.

These fuels vary seasonally and are subject to the particular seasonal mix demanded by the market. The primary use of steam is for pumping and compression.

Just outside of Athens, Greece, at the Motor Oil Refinery in Corinth and the Hellenic Refinery in Aspropyrgos are two plants that have been in operation since 1985 and 1990, with a total of 140,000 and 55,000 operating hours.

Today, more refineries are finding savings in overlooked areas within their processing plants. The effort is focused on how to use by-products. The preferred solution is in-plant cogeneration systems. The refinery cogeneration plants are saving energy and satisfying growing environmental concerns. Refinery requirements for electric power and heat are being met with cogeneration systems consisting of one or more gas turbine-generators and heat recovery steam generators (HRSGs) which utilize the gas turbine waste heat. The generated steam is used either for additional power generation and/or process heat for refinery applications. The refinery must tailor its products to market demand and the gas turbine units must be operated with varying types of fuel, liquid to gaseous, or both simultaneously. The quality, calorific value, viscosity may vary as well as changing electrical and process heat demands. The units can utilize multiple fuels of changing quality and calorific value, including the waste gases derived from the refining process.

Medium size gas turbines are suited for industrial cogeneration, because they are compact, heavy-duty industrial machines with proven reliability and efficiency. The GT10B is rated at 24.6-MW with a simple cycle Heat Rate of 9,970 Btu/kWh. The GT35 is rated at 16.9-MW with a simple cycle heat rate of 10,665 Btu/Wh. Both gas turbines have NO_x emissions of 25 ppm with natural gas fuel.

The refinery near Corinth is about a 1-hour drive north of Athens. The refinery was installed in 1972 and is a medium size refinery with sophisticated conversion; fluid catalytic cracking, alkylation, isomeriza-tion, lube oil processing and traditional distillation, reforming and hydrotreating units. The refinery processes 7,500,000 tons of crude oil per year, and produces 150,000 tons/year of lubrication oil.

The cogeneration plant in the early 1980s had several objectives. One was energy conservation using refinery flare gas (11,000 ton/year). Another was to increase productivity by avoiding shut-downs due to the interruption of electrical power. Others were to increase process steam availability using gas turbine exhaust energy and to reduce operating costs and pollution.

In 1985, the refinery cogeneration plant, consisting of two ABB STAL GT35 Gas Turbine-Generator units, was placed in operation. The two gas turbines exhaust into a single two pressure heat recovery steam generator (HRSG) with supplementary firing capability. The combined electrical output of the two units is 27-MW and a total 52 tons/hour of high pressure steam for power generation and 16 tons/hour of low pressure process steam for refinery purposes.

The primary combustion fuel is refinery flare gas originating from different refinery process streams as waste by-products. It consists mainly of propane and butane in varying proportions ranging from 60% to 100% propane and 40% to 0% butane by volume. The by-product flare gas also contains a varying concentration of H_2S, up to a maximum of 10,000 ppm. Even, 20-25 ppm of H_2S concentration is considered a highly corrosive environment for gas turbine applications, but, the GT35 operates at a low hot blade path temperature. The exhaust temperature is 710°F which is below the melting point of sulfur.

The flare gas varies in qualities, pressures and temperature. Its heating value approximates that of natural gas (1,145 Btu/scf). The cogeneration plant back-up fuel is gasified LPG (liquefied petroleum gas) with a heating value of 2,500 Btu/scf.

To use the flare gas for gas turbine combustion, it is deslugged by a liquid trap which separate the liquid phase from the gas stream. It is then processed through a low pressure compressor and the condensates from the compressed fuel gas are further removed by a gas-liquid separator. Then, the combustion fuel gas is raised to a pressure of 330 psia by a high pressure compressor. The fuel gas is fed to the two gas turbines at a temperature of 203-248°F and pressure of 300 psia.

The Corinth cogeneration plant utilizes its by-product (cost-free) flare gases as the primary combustion fuel for the gas turbines. The cogeneration system eliminates costly refinery downtime due to power outages from the local power grid. The cogeneration plant also eliminates the threat of electrical rate hikes.

The cogeneration plant has reduced NO_x emissions to an air quality well below the acceptable limits imposed by the authorities. The plant's noise emission is 52-dBA at 400 feet.

Utilization of the refinery flare gas has produced an accelerated pay back period for this plant of 2.6 years. The net savings from the plant for 1992 were U.S. $8,000,000.

The Hellenic Refinery in Aspropyrgos, Greece, was installed in 1958, with an initial throughput of 8,500,000 tons per year. This is a state-owned refinery that has been modernized over the years both in capacity output and plant efficiency.

A combined cycle/cogeneration plant was placed in operation in 1990. The cogeneration system consists of two GT10 gas turbine-generator units, two dual pressure heat recovery steam generators (HRSG) of the forced circulation type with by-pass stack, and one ABB condensing steam turbine-generator unit. Saturated steam is also produced for general refinery purposes.

The GT10 unit electrical output is 17-MW and the steam turbine-generator is rated at 15-MW. The combined electrical output of the co-generation system is 49-MW. The generated high pressure steam of 612 psia/760°F from the HRSGs operates the condensing steam turbine-generator set. Each HRSG also produces 18,520 pounds/hour of low pressure steam at 68 psia/342°F for refinery purposes.

The Aspropyrgos cogeneration plant operates normally on refinery flare gas. This fuel has a heating value range of 18,360 Btu/pound to 23,580 Btu/pound, however, heating values of as high as 29,520 Btu/pound have been recorded. The gas turbines are also capable of firing propane, diesel oil and a mixture of the various refinery by-product gas streams. The two turbine units start up on diesel oil.

The refinery installed its own in-plant electrical power and steam plant because of the high price of electrical power and the unreliability of the local electrical power supply. There was also the desire to improve refinery efficiency and plant profitability. The refinery average electrical consumption is 33,300-kW and it exports its excess electrical generation.

Plant reliability has been close to 100% up to its first major inspection. The first major inspection of the two turbines was performed after 25,000 hours of operation (scheduled major inspection is 20,000 operating hours). Unit #2 was inspected and immediately returned by service. More corrosion was found on Unit #1 then expected. It was overhauled, the first two stages of turbine vanes and blades were replaced and cracks in the combustor were field repaired.

Traces of lead, zinc and natrium were found in the gas turbine compressors during the inspection. ABB field service personnel have observed, on occasions, that implosion doors of the air inlet plenums were found to be opened by operators during the plant operation. A metallurgical analysis of the damaged parts indicated that rapid and frequent changes of fuel quality also contributed to damage and cracks in the combustor.

Spare parts and maintenance costs of the two turbines up to the first Major Inspection was U.S. $140,000 for the first 3 years. For a total of 50,000 operating hours at a combined output of 34,000-kW, the service cost comes to less than 0.2 mils/kWhr. The estimated total cost of parts and service of the major inspection of Unit #2 was U.S. $1,120,000, which is equivalent to 2.6 mils/kWhr. The pay back of the plant is estimated at 3.5 years.

In the next several decades natural gas and syngas-fired combined cycle plants are expected to make up as much as 20% of the new electric generating plant additions. Powerplants will utilize techniques such as the atmospheric fluidized bed combustion systems, advanced pressurized fluidized bed combustion systems, integrated gasification-combined-cycle systems, and integrated gasification-fuel cell combined-cycle system.

Many of these will be implemented to facilitate the use of solid fuels such as coal, wood waste and sugar cane waste. The technologies include combined cycle with inlet air cooling, compressor inter-cooling, water or steam injection and the use of water or steam to internally cool turbine airfoils to enhance gas turbine performance and output.

Advances in aircraft engine technology such as airfoil loading, single crystal airfoils, and thermal barrier coatings are being transferred to industrial gas turbines. Improvements in power output and efficiency depends mainly on increases in turbine inlet temperature. The union of ceramics and super alloys provides the material strength and temperature resistance necessary for increased turbine firing temperatures. However, increasing firing temperature also increase emissions.

Catalytic combustors will reduce NO_x formation within the combustion chamber. They will also reduce combustion temperature and extend combustor and turbine parts life.

Increased power in excess of 200+ megawatts can be provided without increasing the size of the gas turbine unit, but the balance of plant equipment such as, pre- and inter-coolers, regenerators, combined cycles and gasifiers will increase the size of the facility needed.

The size of these various process components must be optimized to match each component's cycle with the gas turbine cycle, as a function of ambient conditions and load requirements. Computer systems will interface and control these various processes during steady-state and transient operations.

A major effort will be made to produce gas turbines capable of burning all types of fossil fuel, biomass and waste products. There will be intensified efforts to supply hydrogen, processed from non-fossil resources, for use in petroleum based equipment. The objective will be to produce recoverable, cost effective, and environmentally benign energy.

Ultimately the gas turbine will be needed to burn hydrogen, even with a plentiful supply of fossil fuel. The use of hydrogen eliminates the fuel bound nitrogen that is found in fossil fuels. Hybrid electric cars may use gas turbine generators as range extenders. It would provide a constant power source to continuously charge the on-board battery pack.

With combustion systems currently reducing the dry NO_x level to 25-ppmv and catalytic combustion systems demonstrating their ability to reduce emissions to single digits, the use of hydrogen fuel promises to reduce the emission levels to less than 1-ppmv. The gas turbine will need to achieve this low level in order to compete with the fuel cell.

Hospitals, office complexes, and shopping malls are prime markets for 1 to 5 megawatt powerplants. Operating these small plants, on site, can prove to be more economical than purchasing power through the electric grid from a remote powerplant. This is especially true with the current restructuring of the electric utility industry.

To reduce the expense of turbine units under 5 megawatts, the next breakthrough must be in the production of axial blade and disc assemblies as a single component. The greatest single obstacle to reducing gas turbines costs is in the manufacturing process, machining and assembly. A typical gas turbine has over 4000 parts. About one third of these parts are made from exotic materials with high development costs. Maintaining the turbine is also complicated and requires the same technical skills, as building a new unit.

Advanced design tools such as computational fluid dynamics (CFD) will be used to optimize compressor and turbine aerodynamic design. Producing the blade and disc assembly as a single part would reduce the quantity of parts and the requirements for dimensional tolerances between parts. In very small gas turbines, manufacturers are returning to the centrifugal wheel in both the compressor and the turbine. The advantage is that it can be produced as a single part. This type of small gas turbine generator is now available for the low power market with units from Capstone, Allied Signal, and Elliott. These units operate on gaseous or liquid fuel and generate from 20 to 50 kilowatts.

The gas turbine (turbofan, turbojet, and turboprop) will continue to be the major engine in aircraft applications. In the next decade, over 75,000 gas turbines, turbojet and turbofan, may be built. The gas turbine will also be used to a greater extent in marine applications, mainly in military applications and in fast-ferries. However, the greatest growth will be in land based, stationary, powerplant applications.

References

Boyle, Godfrey, Editor, *Renewable Energy*, Oxford, England: Oxford University Press, 1996.

Fairley, Peter, "Filler Up with Hydrogen," *Technology Review*, Vol. 103 No. 6, November-December, 2000, pp. 56-60.

Giampaolo, Tony, *The Gas Turbine Handbook: Principles and Practices*, Lilburn, GA: The Fairmont Press, Inc., 1997.

Rosenberg, Paul, *The Alternative Energy Handbook*, Lilburn, GA: The Fairmont Press, Inc., 1993.

Spiewak, Scott A. and Larry Weiss, *Cogeneration and Small Power Production Manual*, 5th Edition, Lilburn, GA: The Fairmont Press, Inc., 1997.

CHAPTER 7

LIGHTING UPGRADES

EFFICIENT LIGHTING

There are new ways to reduce energy costs while maintaining quality lighting for different facilities. A large portion of facility utility costs are devoted to lighting. Recent light innovations and energy conservation programs are helping to encourage this task. Lighting upgrades save energy dollars. Lighting systems are often the second greatest use of electric power after heating and cooling systems. With reduced lighting loads, electrical demand savings are often obtained.

Lighting offers several opportunities for energy savings. These include replacing existing lamps with more efficient ones, decreasing the wattage (or brightness), improving controls and changing fixtures and repositioning lights. The lighting energy consumption can be reduced and lower lighting levels decrease air conditioning loads. This can help save energy during warm weather.

Many facilities have outdated lighting components that are nearing the end of their life. Some ballasts may be 20 years old when their life expectancy was 15 years.

Energy-efficient lighting systems in modern office applications strive to optimize lighting quality, maximize energy savings and meet the ergonomic needs of individual workers. The characteristics of the modern office provide an opportunity to optimize office lighting. Optimizing office lighting requires a workspace-specific approach to lighting and control.

In today's energy economy, everyone is seeking technologies or methods to reduce energy expense. Opportunities for lighting retrofits are popular and generally offer an attractive return on investment. Electricity used to operate lighting systems represents a significant portion of total electricity consumed in the United States. Lighting systems consume about 20% of the electricity generated in the United States. Many energy saving technologies are available today.

Technological advances include lamps, ballasts, and lighting controls, such as improved metal halide systems with pulse start ballasts. Metal halide lamps are High Intensity Discharge (HID) lamps that produce an intense white light. These lamps were designed in the 1960s for stadium lighting and other area applications. They are available in sizes ranging from 32 watts to 2000 watts.

Metal halide lighting applications are the fastest growing segment of the lighting market. Metal halide lamps are useful for lighting large areas efficiently. They are used in manufacturing areas, high bay areas, warehouses, parking lots and other commercial and industrial spaces. Their efficacy in lumens per watt is excellent and they provide a white light with good color rendering along with compact size and long life (Figure 7-1). Metal halide lamps are a type of gas discharge lamp and like mercury vapor, fluorescent, high pressure sodium and neon lights, they use a ballast to start the discharge in the tube.

Lighting energy management has three basic tasks:

- Identify the light quantity and quality needed to perform the visual tasks,
- Increase the light source efficiency, and
- Optimize lighting controls. A lighting upgrade produces a better environment and it can also pay for itself through energy and maintenance savings.

LOAD MANAGEMENT

An important opportunity to decrease energy consumption is by controlling the demand for electricity at a particular time. This is called load management. It can allow an electric utility to offer lower electric rates at particular times of the day and year.

Time-of-use rate structures (TOURS) are used by some utilities. This is likely to become more widespread as utilities seek to smooth out daily load patterns. The minimum level of demand day and night, season after season is known as the base load. At certain times during the day and during the year the demand rises above the base load and electricity usage reaches a maximum. This occurs when people return home from work and turn on their appliances or on very hot days, when many air conditioners and fans are operating during the same time period. This type of demand is called a peak load.

Different types of powerplants are used to provide base and peak load power. Base load plants operate almost continuously because they are the least expensive to operate or because they are more difficult to start up and shut down. Nuclear powerplants are almost always used as base load plants. Peaking plants generally burn oil or gas since such plants are fairly easy to start up and shut down.

Base load plants are paying for their way around the clock, while peaking plants must be paid for even when they do not operate. Base load operation is usually less expensive than peaking operation and therefore generating costs to the utility are less during off-peak periods than during on-peak periods.

Utilities usually keep a certain amount of generating capacity in reserve in order to allow maintenance of plants. If the peak demand increases and eliminates this reserve, it is necessary for the utility to build new powerplants. If the utility can shift more demand to base load periods, the construction

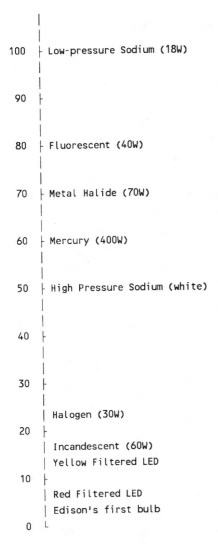

Figure 7-1. Performance (lumens per watt) of different kinds of lights

of new plants can be minimized. Peak demand hours generally occur in the late afternoon and early evening, and during summer mid-afternoons due to air conditioning loads.

A load that normally runs continuously or at random times during the day can be wired to a clock timer that will switch the load off during the peak demand period.

TOURS rates may vary either during the day or over the seasons, or both. With both daily and seasonal TOURS rates in effect, the most expensive time to operate a load would be on a hot summer afternoon.

Energy storage devices can store heat or cold during off-peak hours, when electricity is cheaper, and take it out of storage for use during peak demand periods. Off-peak air conditioning can use less energy than normal daytime operation, since the compressor and heat exchangers in the unit will be operating in nighttime temperatures, which are generally lower than those during the day.

Heat storage devices include oil or magnesia brick storage heaters. These may be substituted for conventional electric resistance baseboard units. Air conditioning storage devices include chilled water tanks, cooled rock beds and ice makers combined with storage tanks.

The paybacks for load management tend to be greater for large power users. Large consumers may already pay demand charges based on monthly peak demand. When demand charges are high, it can quickly pay to shift or change some energy use or to store energy in order to reduce peak demand.

If the local electric utility offers time-of-use rates, you can reduce costs shifting loads to off-peak hours. On-peak rates can be four times off-peak costs per kilowatt-hour.

HVAC EFFECTS

Most of the energy consumed by lighting systems is converted to light, heat and noise, which dissipates into the building. If the amount of energy consumed by a lighting system is reduced, the amount of heat energy losses into the building will also be reduced, and less air-conditioning will be required. However, the amount of wintertime heating may need to be increased to compensate for a lighting system that dissipates less heat.

Since most offices use air-conditioning for more months per year than heating, a more efficient lighting system can significantly reduce air-conditioning costs. Air-conditioning is usually more expensive that heating and the air-conditioning electricity savings are usually more than the additional heating costs.

HUMAN ASPECTS

Lighting retrofits improve the lighting quality and the performance of workers. Recent advances in technology offer an opportunity for energy conservation and improve worker productivity. High frequency electronic ballasts and tri-phosphor lamps provide improved CRI, less audible noise and lamp flicker. These benefits can improve worker productivity and reduce headaches, fatigue and absenteeism.

LIGHTING MANAGEMENT AND AUDITS

The goal of lighting management is to achieve the right quantity and quality of lighting in any given facility. Today, special emphasis is placed on achieving this goal at the lowest cost possible to the facility owner.

Lighting management recommendations are published by the Illuminating Engineering Society (IES) (Table 7-1). These are general guidelines. Since each lighted space has its own unique lighting requirements, IES recommendations are augmented by an analysis of the lighted space. This analysis is known as a lighting audit.

The lighting audit studies the current lighting system to assess the type and number of fixtures in place. The audit will also note how much natural daylight is available. In evaluating which new components should be used to retrofit fixtures, the audit must also consider any special wiring requirements that must be followed or corrected.

The audit also looks for opportunities to increase the energy-saving potential of the system. For example, the audit might recommend installing occupancy sensors that turn lights on or off automatically, or reflec-

Table 7-1. Recommended Light Levels in Foot-candles

Commercial interiors	
Art galleries	30-100
Banks	50-150
Hotels (rooms and lobbies)	10-50
Offices	30-100
• Document handling	50-75
• Hallways	10-20
• Rooms with computers	20-50
Restaurants (dining areas)	20-50
Stores (general)	20-50
Merchandise	100-200
Institution Interiors	
Auditoriums/assembly places	15-30
Hospitals (general areas)	10-15
Labs/treatment areas	50-100
Libraries	30-100
Schools	30-150
Industrial Interiors	
Ordinary tasks	50
Stockroom storage	30
Loading and unloading	20
Critical tasks	100
Highly critical tasks	200-1000
Exterior	
Building security/Parking	1-5
Floodlighting (high/low)	5-30

tors that increase light levels and could permit the removal of one or two lamps from four-lamp fixtures.

The completed lighting audit should specify the most appropriate energy-saving technologies which could be used to upgrade the lighting system.

LIGHTING AUDITS

Lighting audits involve the measurement of lighting intensity and the wattage in foot-candles (f.c.). This is compared to the recommended values and followed by actions to modify the lighting system. An audit may show that some parts of the building are overlit. Lights can be replaced by bulbs of lower wattage if the lighting audit shows an overlit space.

Some states require a lighting audit for buildings of a certain floor area. Massachusetts requires an audit of buildings with floor areas of 10,000 square feet or more. The results of the audit must be filed with the State Building Code Commission.

Existing lights can be replaced by bulbs of equal brightness but lower wattage. This usually means replacing incandescent bulbs by fluorescent ones, which can result in energy savings of 60% or more. Replacement by fluorescents normally required installation of a new fixture, but newer fluorescent bulbs that screw into standard light sockets are now available.

Existing fluorescents can be replaced by more efficient ones, saving 15 to 20% in energy consumption. The fluorescent ballast transformer may have to be changed.

LIGHT LEVELS

If you can turn off a lighting system for the majority of time, the expense to upgrade lighting sources may not be justified. For many years, lighting systems were designed with the notion that no space can be over-illuminated.

Starting in 1972 IES recommended light levels declined by 15% in hospitals, 17% in schools, 21% in office buildings and 34% in retail buildings. An excessively illuminated space results in unnecessary energy consumption. Excessive illumination is not only wasteful, but it can reduce the comfort of the visual environment and decrease worker productivity. Table 7-2 lists some areas of commercial buildings along with recommended light levels, CRI, color temperature and glare.

The relocation of light fixtures can often improve the quality of

Table 7-2. Lighting for a Commercial Building

Area	Light Level	CRI	Color Temperature	Glare
Executive Office				
General	100fc	≥80	3000K	VCP≥70
Task	≥50fc	≥80	3000K	VCP≥70
Private Office				
General	30-50fc	≥70	3000-3500K	VCP≥70
Task	≥50fc	≥70	3000-3500K	VCP≥70
Open Office				
General	30-50fc	≥70	3000-3500K	VCP≥90
Task	≥50fc	≥70	3000-3500K	VCP≥90
Hallways				
General	10-20fc	≥70	3000-3500K	VCP≥70
Task	10-20fc	≥70	3000-3500K	VCP≥70
Reception/Lobby				
General	20-50fc	≥80	3000-5000K	VCP≥90
Task	≥50fc	≥80	3000-3500K	VCP≥90
Conference				
General	10-70fc	≥80	3000-4100K	VCP≥90
Task	10-70fc	≥80	3000-4100K	VCP≥90
Drafting				
General	70-100fc	≥70	4100-5000K	VCP≥90
Task	100-150fc	≥780	4100-5000K	VCP≥90

light in a space and decrease the required brightness. This may involve bouncing light walls or other surfaces as well as bringing the light closer to the task. Movable fixtures can produce similar effects. Timing controls can automatically turn lights on or off at specified times.

Even greater energy savings can be realized by using photocell controls for lighting circuits. These controls turn lights off or on when daylight entering the photocell passes through a preset level. The payback for these controls is usually 3 to 6 years. A manual dimmer switch allows the user to control the brightness and minimize the energy consumption of a particular lighting fixture.

In large buildings, placing lights in some areas on a centralized load controller, which switches them off when building electricity demand exceeds a certain level, can save on demand charges by reducing the peak load. Demand charges are paid by large electricity consumers and represent the utility's cost of keeping the generating capacity available to meet the user's peak demand.

SOURCE EFFICACY

Increasing the source efficacy of a lighting system means replacing or modifying the lamps, ballasts and/or luminaries to become more efficient. Due to the inter-relationships between the components of current lighting systems, ballast and luminaire retrofits should also be considered source upgrades. Increasing the efficacy means getting more lumens per watt out of the lighting system.

Efficacy describes an output/input ratio, the higher the output for a given input, the greater the efficacy. Efficacy is the amount of lumens per watt from a source. Light sources with high efficacy can provide more light with the same amount of power (watts), when compared to light sources with low efficacy.

A magnetic ballast and T12 lamps could be replaced with T8 lamps and an electronic ballast, which is a more efficient system. Another retrofit that would increase source efficacy is to improve the luminaire efficiency by installing reflectors and more efficient lenses. This retrofit would increase the lumens per watt, since the reflectors and lenses allow more lumens to escape the luminaire, while the power input remains constant.

Energy-efficient lighting savings depend on the following basic energy equation:

Energy (kilowatt-hours) = Power (kilowatts) × Time (hours)

This equation tells us that energy-efficient lighting can save on lighting system operating costs in either of two ways:

- Reducing the power demand of the lighting system, or
- Reducing the system's hours of use.

Both methods are used in today's energy-efficient lighting technologies.

The use of T8 fluorescent lamps with its 32-watt rating consumes 20% less energy than a standard 40-watt T-12. Its smaller diameter also allows more of the T8's light output to escape from fixtures and reach the workplace below. Standard fluorescents also use a single phosphor coating to convert ultraviolet light produced inside the lamp into visible light as it leaves the lamp. The T8 incorporates three rare-earth phosphors, giving it superior color characteristics. The more natural light produced by the T8 helps to reveal the true colors of objects. This allows occupants to experience less eye strain, perform tasks more efficiently and make fewer errors.

ELECTRONIC BALLASTS

Electronic ballasts are another new technology. The fluorescent lamp requires a ballast to light the lamp and regulate its operation. Standard ballasts use a transformer based magnetic coil technology, while the latest innovation in ballast design is the electronic ballast.

The electronic ballast operates at higher frequencies, about 20,000-Hz compared to 60-Hz for magnetic ballasts. These higher frequencies result in energy savings of up to 40%, extend rated ballast life to 20 years, and virtually eliminate any flicker. Since electronic ballasts produce less heat, using them in a lighting upgrade can lower a facility's air-cooling requirements.

Luminaires are the glass or metal fixtures that reflect or diffuse light from the bulb. These fixtures generally trap light and also spread it in particular directions. By using a different luminaire you can reduce energy consumption by up to 40%. Manufacturer's literature can provide information on the effectiveness of a specific fixture.

Specular reflectors feature a mirror-like surface. They are mounted inside existing fluorescent fixtures to direct light more efficiently onto the workplace. They allow up to half the lamps in each fixture to be removed providing up to 50% energy savings.

Energy costs are frequently divided between demand charges (kW) and consumption charges (kWh). Most lighting retrofits can provide savings in both demand and consumption charges.

The potential for lighting retrofits also includes improving the visual surroundings and worker productivity. If a lighting retrofit reduces lighting quality, worker productivity can drop and the energy savings may be offset by reduced profits. This occurred in the lighting retrofits of the 1970s, when workplace levels dropped due to extensive delamping initiatives. Since then however, there have been major advances in technologies and today's lighting retrofits can reduce energy expenses while improving lighting quality and worker productivity. An understanding of lighting fundamentals is required to ensure that lighting upgrades will not compromise visual comfort.

LIGHTING QUANTITY

Lighting quantity is the amount of light provided to a space. Unlike light quality, light quantity is easy to measure and describe. Lighting quantity is affiliated with watts, lumens and foot-candles. The watt is the unit of electrical power. It defines the energy used by an electrical appliance. The amount of watts consumed represents the electrical input to the lighting system.

The output of a lamp is measured in lumens. The amount of lumens represents the brightness of the lamp. A four-foot fluorescent lamp provides about 2,500 lumens. The number of lumens describes how much light is being produced by the lighting system.

The number of foot-candles will show how much light is actually reaching the workspace. Foot-candles result from watts being converted to lumens. The lumens escape from the fixture and travel through the air to reach the workspace. You can measure the amount of foot-candles with a light meter.

Foot-candle measurements express the aftermath and not the effort of a lighting system. The Illuminating Engineering Society (IES) recommends light levels for specific tasks using foot-candles.

IES LIGHT LEVELS

The Illuminating Engineer Society (IES) is the largest group of lighting professionals in the United States. Since 1915, IES has specified light levels for various tasks. The light levels recommended by the IES usually increased until the 1970s.

More recently, more effort has been placed on occupant comfort which decreases when a space has too much light. Several experiments have shown that some of IES's light levels were exorbitant and worker productivity was affected due to visual discomfort. As a result of these findings, IES revised many of their light levels downward.

The tradition of excessive illumination continues in many office areas. In the past, lighting designers would identify the task that required the most light and design the lighting system to provide that level of illumination for the entire space.

In an office, there are several tasks including reading information on computer screens. Too much light tends to reduce screen readability. In an office with computers, there should be up to 30 foot-candles for ambient lighting. Small task lights on desks can provide the additional foot-candles needed to achieve a total illuminance of 50 to 75 foot-candles for reading and writing.

LIGHTING QUALITY

Lighting quality is more subjective and has a major influence on the attitudes and actions of occupants. Different environments are created by a lighting system. Occupants should be able to see clearly without glare, excessive shadows or any other uncomfortable appearance of the illuminated area.

Although occupant behavior also depends on interior design and other factors, lighting quality has a critical influence. Lighting quality depends on uniformity, glare, color rendering index and coordinated color temperature. Improvements in lighting quality can yield high dividends for businesses since gains in worker productivity often result when lighting quality is improved.

UNIFORMITY

The uniformity of illuminance describes how evenly the light spreads over an area. Uniform illumination requires the proper luminaire spacing. Non-uniform illuminance can create bright and dark spots.

In the past, light designers have stressed uniform illumination. This eliminates any problems with non-uniform illumination and makes it easy for workers to move around in the work environment. But, uniform lighting applied over large areas can waste large amounts of energy. In a large area 20% of the floor space may require high levels of illumination (100 foot-candles) for specific visual tasks. The remaining 80% of the area may only require 40 foot-candles. Uniform illumination over the entire space would require 100 foot-candles over the entire area.

GLARE

Glare is caused by relatively bright objects in an occupant's field of view. Glare is most likely when bright objects are located in front of dark environments. Contrast is the relationship between the brightness of an object and its background. Most visual tasks become easier with increased contrast, but too much brightness causes glare and makes the visual task more difficult. Glare is a serious concern since it usually causes discomfort and reduces worker productivity. The quality of light can be specified by the CRL, CCT and VCP.

VISUAL COMFORT PROBABILITY

The visual comfort probability (VCP) is a rating given to a luminaire which indicates the percent of people who are comfortable with the glare. A luminaire with a VCP of 80 means that 80% of occupants are comfortable with the amount of glare from that luminaire. A minimum VCP of 70 is recommended for general interior spaces, while luminaires with VCPs above 80 are recommended in computer areas and other work intensive environments. High-glare environments are characterized by excessive illumination or reflection. Often these bright areas are found around luminaires.

Reducing glare can be achieved by using indirect lighting, deep cell parabolic troffers or special lenses. Although these measures reduce glare, luminaire efficiency is decreased since more light is trapped in the luminaire. A better solution is to minimize glare by reducing ambient light levels and using task lighting techniques.

Today's office environment includes the use of computer monitors which are vulnerable to glare and discomfort. When reflections of ceiling lights are visible on the screen, the user has difficulty reading the screen. This is called discomfort glare and is common in rooms that are uniformly illuminated by luminaires with a low VCP. Moveable task lights and luminaries with high VCPs should be used in these environments.

COLOR INDEX AND TEMPERATURE

Light sources are specified based on two color-related systems: the Color Rendering Index (CRI) and the Coordinated Color Temperature (CCT). The CRI provides a measure of how colors appear under a given light source. The index ranges from 0 to 100 and the higher the number is the easier it is to distinguish colors. Sources with a CRI greater than 75 provide good color rendition and sources with a CRI of less than 55 provide poor color rendition. Offices illuminated by T12 Cool White lamps have a CRI of 62.

A light source with a high CRI is needed for visual tasks that require the occupant to distinguish colors. An area with regular color printing requires illumination with excellent color rendition. Outdoor security lighting need not have as high a CRI, but a large quantity of light is generally required.

The coordinated color temperature (CCT) describes the color of the light source. The color differences are indicated by a temperature scale. The CCT is measured in degrees Kelvin and approximates the color that a black-body object would radiate at a certain temperature. Red light has a CCT of 2000K, white light a CCT of 5000k and blue light a CCT of 8000K.

Traditionally, office environments have been illuminated by Cool White lamps, which have a CCT of 4100K. A more recent trend is to use 3500K tri-phosphor lamps, which provide a more neutral light color.

SENSORS AND CONTROLS

Occupancy sensors can reduce lighting system energy consumption by 15 to 20%. Eliminating the unnecessary use of lighting, they act as motion sensors and are programmed to automatically turn off the lights in a room or space a specified period of time after all the occupants have left.

Passive or active sensing is used. A passive unit uses infrared sensors that respond to changes in object surface temperature and movement. An active unit sends out either an ultrasonic or microwave signal and then responds to changes in the signal as it returns to the receiver at the source.

The device operates as a normal, single-pole light switch with an infrared sensor, which senses whether or not the room is occupied. If the infrared sensor detects that the room is unoccupied, the switch will open, turning the lights off. This type of switch can be used to control areas of up to about 400 square feet.

For outdoor or exterior lighting, timer or photoelectric switches can be used to turn the lights on when needed and off when not needed. The time clock needs to keep up with time changes and the different seasons. A photoelectric switch will keep the lights turned on whenever it is dark outside, which is often wasteful.

By combining a photoelectric switch with a time clock, the lights can be set to come on whenever it is dark, except during certain hours. The time clock could be set to give power to the lights from 7:00 a.m. to 10:00 p.m. between these hours, power is fed to the outdoor lights. But, the photoelectric switch is also wired into the circuit and will not allow any current to flow through to the lights until it senses darkness. This permits the lighting to turn on not only during nighttime hours, but also during storms when the sky may become darkened and outdoor lighting might be desired.

Dimmer units allow the light output from fixtures to be adjusted up or down to meet the requirements of a given space. Some dimmers offer manual control while others incorporate photosensors, allowing the dimmers to respond automatically to changes in ambient lighting conditions.

The light-sensitive dimmers maintain a constant, preset level of light by raising or lowering fixture light output as needed. Dimming

control systems can also be included in a lighting upgrade, since electronic ballasts have some dimming capability.

Modern controls can turn entire systems off when they are not
needed. This allows energy savings to accrue quickly. The Electric Power
Research Institute (EPRI) reports that spaces in an average office building may only be occupied 6-75% of the time.

In areas that have sufficient daylight, dimmable ballasts can be used
with control circuitry to reduce energy consumption during peak periods.
Although there may be some shedding of lighting load along the perim-
eter of the space, these energy cost savings may not represent a large part
of the building's total lighting load. Applications of specialized technologies such as dimmable ballasts can be dispersed in several buildings.

TASK LIGHTING

Task lighting involves improving the efficiency of lighting by providing the appropriate illumination for each task. This often results in a
reduction of ambient light levels, while maintaining or increasing the
light levels on a particular task. The light level needed on a desk may be
75 foot-candles while the light needed in wall aisles is only 20 foot-
candles. Uniform lighting would create a workplace where the ambient
lighting provides 75 foot-candles throughout the entire workspace. Task
lighting would create an environment where each desk is illuminated to
75 foot-candles, and the aisles only to 20 foot-candles. Task lighting is
especially qualified for office environments with computers and modular
furniture that can incorporate task lighting under shelves. Moveable desk
lamps can also be used for task illumination.

In work spaces where a variety of visual tasks are performed, each
employee has optimal lighting preferences. Most workers prefer lighting
systems designed with task lighting because it is flexible and allows
individual control.

Task lighting techniques are also suitable in industrial facilities
where the ambient lighting can be reduced by task lighting. Task lights
can also be installed on fork-lift trucks to supplement the headlights for
use in warehouses.

INCANDESCENT LAMPS

The oldest electric light technology is the incandescent lamp. Incandescent lamps also have the lowest lumens per watt and have the shortest life (See Table 7-3). They produce light by passing a current through a tungsten filament, causing it to become hot and glow. As the tungsten emits light, it progressively evaporates, eventually causing the filament to weaken and break open.

Table 7-3. Lamp Characteristics

Wattage\ Life (hours)	Lumen Maintenance	Color Rendition	Light Direction Control	Relight Time	Fixture Cost
Incandescent 15-1500W 750-12,000	Fair to Excellent	Excellent	Very good, Excellent	Immediate	Low
Fluorescent 15-219W 7,500-24,000	Fair to Excellent	Good to Excellent	Fair	Immediate	Moderate
Compact Fluorescent 4-40W 10,000-20,000	Fair	Good to Excellent	Fair	Immediate- 3 seconds	Moderate
Mercury Vapor 40-1000W 15,000-25,000	Very good	Poor to Excellent	Very good	3-10 Minutes	High Moderate
Metal Halide 175-1000W 1,500-15,000	Good	Very good	Very good	10-20 Minutes	High
High-pressure Sodium 70-1000 10,000-24,000	Excellent	Fair	Very good	Less than 1 minute	High
Low-pressure Sodium 35-180W 18,000	Excellent	Poor	Fair	Immediate	High

Incandescent lighting consists of common light bulbs, certain types of floodlights, (quartz or tungsten-halogen) and most high-intensity reading lights. Incandescent lighting is the least expensive bulb to purchase and produces a soft light. Incandescent lighting gives far less light per energy dollar than other types of lighting. Incandescent bulbs do not last as long as other types of lighting. A tungsten alloy is used as the filament and the glass bulb maintains a vacuum for the glowing filament. Without the vacuum, the tungsten filament would quickly burn up.

Halogen lamps are a type of high-efficiency incandescent lamp. The lamp uses halogens to reduce evaporation of the tungsten filament. Halogen lamps provide a 40 to 60% increase in efficiency with a whiter, brighter light than standard incandescents.

Special types of incandescent lamps for flood lighting include quartz and tungsten-halogen units that operate on the same principle as regular incandescent bulbs but are constructed of slightly different materials to give longer service and to provide more light.

One product that is used with incandescent light bulbs is a small disk which is inserted into the socket before the light bulb is screwed into place. These devices are variable electrical resistors which limit the initial rush of current to the light bulb. By limiting the initial rush of current, the temperature of the bulb's filament rises relatively slowly compared to the temperature rise without the resistor. This reduces the amount of shock which the filament must endure and extends the life of the bulb. With this shock reduced, the bulbs provide service for a longer time. The light bulb uses the same amount of electricity except for the initial inrush.

Another device used for extending lamp (bulb) life is a diode, which cuts off half of the power to the bulb. The bulb will last longer but emit half the light. This device is useful in locations that are difficult to reach. No energy is saved for the amount of lumens emitted, but there will be fewer light bulbs to purchase. Another technique is to use bulbs which are rated at 130 volts, rather than the normal 120 volts. The 130 volt bulbs have a stronger filament, which will last longer and the price differential is not too large.

Incandescent lamps are sold in large quantities because of several factors. Many fixtures are designed for incandescent bulbs and lamp manufacturers continue to market incandescent bulbs since they are easily produced. Consumers purchase incandescent bulbs because they have low initial costs, but if life-cycle costs are considered, incandescent lamps are usually more expensive than other lighting systems with higher efficacies.

COMPACT FLUORESCENT LAMPS

Compact Fluorescent Lamps (CFLs) are more energy efficient, longer lasting replacements for incandescent lamps. Compact fluorescent lamps are miniature ballast-lamp systems designed as a direct replacement for incandescent lamps. They provide similar light quantity and quality while requiring about 30% of the energy of incandescent lamps. They also last up to 10 times longer than incandescent units.

A CFL provides energy savings of 60 to 75%, with only a slight decrease in light levels. The typical CFL has a rated life of 10,000 hours which is about 12 times the life of a standard incandescent bulb. Although it has a higher initial price, a CFL can pay for itself in 1 to 2 years through energy and maintenance savings.

CFLs are available in several styles and sizes. They are available as self-contained units or as discrete lamps and ballasts. A self-contained unit screws in to the existing incandescent bulb socket.

There are short tubular lamps called PL or SLS lamps that can last more than 8,000 hours. The ballasts at the base of the tube can last longer than 60,000 hours.

Self-contained CFLs can be used to replace incandescent lamps in table lamps, downlights, surface lights, pendant luminaires, task lights, wall sconces, and floor lights. The size of CFLs has decreased recently allowing more applications for retrofits. Many fixtures are specially designed for CFLs in new facilities.

Improvements to CFLs have been occurring each year since they became available. In comparison to the first generation of CFLs, today's products provide 20% higher efficacies as well as instant starting, reduced lamp flicker, quiet operation, smaller size and lighter weight. Dimmable CFLs are also available.

Since CFLs are not point sources like incandescent lamps, they are not as effective in projecting light over some distance. The light is more diffused and harder to focus on some target in directional lighting applications. Like most fluorescent lamps, CFLs do not last as long when switched on and off and the lumen output decreases over time. The light output of CFLs can also be reduced when used in luminaires that trap heat near the lamp.

CFL ballasts include magnetic units with a one year warranty and electronic or hybrid units with up to a 10 year warranty. Some units may have difficulty starting when the temperature drops below 40°F, while others will start at temperatures below freezing. Some CFL units can produce an excess of 110% total harmonic distortion (THD), while others produce less than 20% THD.

FLUORESCENT LIGHTING

Fluorescent lighting systems use long, cylindrical lamps and emit light from a phosphorescent coating on the surface of the lamp. A ballast transformer is used to increase the voltage high enough to arc from one end of the lamp to the other. This emits electrons, which create light when they strike the coating on the lamp's surface. Fluorescent lighting provides several times as much light per energy dollar (lumens per watt) as incandescent lighting. Fluorescent lighting provides between four and six times the light per dollar of incandescent lighting.

Fluorescent lighting flicker is caused by the alternating current that is used. Since the voltage alternates between positive and negative and goes through zero for an instant the arc within the fluorescent tube is momentarily extinguished. This occurs 120 times per second and may be noticeable. Fluorescent lighting can sometimes cause people's eyes to become tired or irritated. This problem is usually noticed when fluorescent lighting is the only type of lighting in a certain area. In areas where there is an additional light source such as incandescent lighting or sunlight, there is rarely a problem, since the other sources provide enough background light to swamp the flicker effect.

Another consideration for fluorescent lighting is the color of the light. Incandescent lighting has a slightly yellow tone. The most common type of fluorescent lamp is the cool white lamp, with a slightly blue tone.

Fluorescent lighting manufacturers have developed different types of fluorescent lamps which produce different shades of light. Warm white and incandescent white fluorescent lamps are available. Fluorescent tubes last several times longer than incandescent lamps, which helps to offset their initially higher cost. This also makes them easier to maintain, since the lamps need to be changed less often.

Older fluorescent fixtures may develop a buzzing sound. In high quality fixtures this is not typically a problem, since they are made to have a very low noise level. The newer electronic ballasts eliminate the noise from transformer humming ballasts.

Fluorescent lamps are relatively efficient, have long lamp lives and are available in a variety of styles. The lumens per watt can exceed 70 and the rated life can exceed 20,000 hours for standard units and 40,000 hours for special long life systems. A common fluorescent lamp that has been used in offices is the four-foot F40T12 lamp used with a magnetic ballast. These lamps are being replaced by F32T8 and F40T10 tri-phosphor lamps with electronic ballasts.

In an F40T12 lamp, the F stands for fluorescent, 40 stands for 40 watts and the T12 refers to the tube thickness (diameter). The tube thickness is measured in 1/8 inch increments so a T12 is 12/8 or 1.5 inches in diameter. A T8 lamp is 1 inch in diameter.

Some lamps include additional information, such as the CRI or CCT. Usually, CRI is indicated with one digit, an 8 means the CRI is 80. CCT is indicated by the two digits following this, 35 refers to 3,500K. F32T8/841 indicates a 32 watt, 1-inch-diameter lamp with a CRI = 80 and a CCT = 4,100K. Some manufacturers use letter codes referring to a lamp color. CW means Cool White with a CCT of 4,100K.

Lamps with ES, EE or EW indicate that the lamp is an energy-saving type. These lamps consume less energy than standard lamps, but they also produce less light.

Tri-phosphor lamps use a coating on the inside of the lamp to improve performance by about 30%. These lamps usually provide about a 25% higher color rendition. A bi-phosphor lamp like the T12 Cool White has a CRI of 62. Tri-phosphor lamps have a CRI of over 80 and allow occupants to distinguish colors better. Tri-phosphor lamps are usually specified with T8 or T10 systems and electronic ballasts. Lamp flicker and ballast humming are considerably reduced with electronic ballasts.

Fluorescent fixtures are usually best for lighting a large area, and incandescents are often best for lighting specific areas that need light only sporadically. The new, small fluorescent lamps that are shaped similarly to an incandescent bulb and screw into a typical socket are a good energy saver. They do cost several times as much as incandescent bulbs, but they will last several times longer and use only a fraction of the electricity that a regular bulb will use. These small fluorescents will save money over the long term.

HID LIGHTING

High intensity discharge (HID) lighting is used mostly for outdoor lighting using floodlights. HID lighting gives an extremely bright and concentrated light.

HID lights include mercury vapor, metal halide, and high pressure sodium types. These lights use the same operating principles as fluorescent lighting, but are constructed with different types of materials and methods.

They produce light by discharging an electric arc through a tube filled with gas, but generate more light, heat and pressure in the arc tube than fluorescent lamps and have an initial high intensity discharge. Like incandescent lamps, HIDs have a small light source, that approximates a point source. This means that reflectors, refractors and light pipes can be used to direct the light. HID lights last a little longer than fluorescents typically reaching 16,000 to 24,000 hours.

Metal halide lamps are similar to mercury vapor lamps, but they provide substantially more light and give off a much clearer light than mercury vapor lamps. Metal halide lamps do not last as long as the mercury vapor lamps (16,000 to 20,000 hours), but the greater light output and the better quality light make up for the shorter life span.

The most efficient HID lighting system is the high pressure sodium system. These operate on the same principle as the others, but give more light per energy dollar, have a life of up to 24,000 hours. They provide an efficacy of up to 140 lumens/W.

One problem with high pressure sodium lighting is the noticeable orange-tinted light. An improved color version is available, but it has only about 1/2 the life and about 15% less efficacy than standard bulbs. High pressure sodium lighting is used primarily in locations where color rendering is not important such as street, parking areas and walkways. HID lighting fixtures are also expensive to purchase.

HIDs were originally developed for outdoor and industrial applications, but they are also used for office, retail and other indoor applications. Most HIDs need time to warm up and should not be turned on and off for short intervals. The relight time ranges from less than one minute for high pressure sodium, to up to 20 minutes for metal halide. Mercury vapor relight times are about 1/2 of that of metal halide. Besides mercury vapor, metal halide and high pressure sodium, low pressure sodium is also used.

MERCURY VAPOR LAMPS

Mercury vapor was the first type of HIDs. Today they are relatively inefficient, provide poor CRI and have the most rapid lumen depreciation rate of all HIDs. More cost-effective HID sources have replaced mercury vapor lamps in most applications. Mercury vapor lamps provide a white-colored light with a slight blue tint, and its efficiency is similar to fluorescent bulbs. Most mercury vapor lamps are rated at about 20,000 hours. The efficacy is about 60 lumens/W. Mercury vapor lighting has been largely replaced by metal halide lighting. A popular lighting upgrade has been to replace mercury vapor systems with metal halide or high pressure sodium systems.

METAL HALIDE LAMPS

Metal halide lamps are similar to mercury vapor lamps, but contain different metals in the arc tube, providing more lumens per watt with improved color rendition and improved lumen maintenance. They have almost twice the efficacy of mercury vapor lamps and provide a white light for industrial facilities, sports arenas and other spaces where good color rendition is required.

Metal halide lamps are often used to retrofit incandescent, fluorescent and high pressure sodium lighting systems. Metal halide lamps produce light 3 to 5 times more efficiently than incandescent lamps. For an equivalent amount of energy, metal halide lamps produce five times the amount of light without the heat associated with incandescent lamps.

They emit a natural white light rather than the yellow light of high pressure sodium. Metal halide lamps are relatively unaffected by changes in ambient temperature.

Sometimes the upgrade can be accomplished using the existing fixture by adding a ballast, a new lamp and a replacement socket. This can be more cost effective than whole fixture replacement.

The arc tube is the core of a metal halide lamp. A 32-watt arc tube has a pea-sized chamber that emits as much light as a 4-foot-long fluorescent lamp. A 2000 watt spot light has an arc tube the size of a two cell flashlight enclosed in an outer bulb the size of a soccer ball. This lamp emits almost as much light as 120 standard 100-watt incandescent bulbs.

The arc tube is made of fused silica (quartz). Although it resembles glass, quartz has a very high melting point. The walls of this chamber typically reach about 1000°C in an operating arc tube. At the center of the discharge itself, the temperature can be as high as 5000°C, providing an intense and efficient source of light.

When the arc tube is started, the heat from the electric arc discharge heats up the chamber and causes the mercury in the arc tube to vaporize, increasing the pressure. This is called the warm up phase. When all the mercury is vaporized, the lamp is fully warmed up.

An operating arc tube contains mercury vapor at a pressure of 4 to 20 atmospheres with a small partial pressure of metal halide molecules. The metal atoms are responsible for most of the emitted light.

Various chemicals are added to obtain cool white, warm white or daylight type lamps. A small amount of a starting gas (argon or xenon) is also used in the chamber.

Ballasts which operate metal halide lamps were originally designed in the 1960s and 1970s. Design changes were limited because new designs needed to be compatible with existing lamps. Lamp designs had only minor improvements due to the need for new lamps to operate on existing ballasts. Several new wattages have been added, but the lamps have changed very little over the past 30 years.

The role of the ballast is first to initiate the discharge and then to maintain a stable flow of electric current. The current heats the arc tube and ensures that sufficient metal halide molecules are vaporized. The mechanism of light emission is different from the light that is emitted from a hot filament. Unlike a filament lamp which produces a continuous spectra, a metal halide lamp provides a line spectrum, emitting light in many discrete regions of the visible spectrum. This direct excitation of atoms to emit visible light results in the high efficiency of these lamps, and substantially less heat is generated than in an incandescent filament lamp of the same wattage.

PULSE START SYSTEMS

The newer technology includes lamps and ballasts that are designed together with interactive features. This trend started with low wattage lamps introduced in the 1980s and early 1990s. Now all major manufacturers provide high wattage lamps of 150 to 450 watts and compatible ballasts in a package that uses pulse start technology. Pulse start systems are attractive for indoor lighting for commercial and industrial areas. The wattage range matches the lighting levels needed for indoor spaces.

This was a major step for metal halide lighting. Pulse start technology is expected to replace standard metal halide lamps in 175, 250, 400 and 1000 watt applications which are about 70% of the total market.

In metal halide lamps a critical area was the argon gas fill pressure in the arc tube chamber. Too low a pressure creates a build up of tungsten that causes a light-absorbing film on the arc tube wall. Higher pressures reduce this build up but make starting more difficult.

In the late 1960s and early 1970s ballasts used higher peak voltages to improve metal halide lamp starting. This higher voltage would start the arc tube with higher fill pressures. Standard metal halide ballasts use this higher voltage. Since it is in effect during regular operation, it causes electrode damage and reduces the light output over a period of time.

Pulse start systems use high voltage ignitors to initiate the electric discharge. The older metal halide ballasts used a third electrode in the arc tube called a starter electrode. The newer electronic pulse will start arc tubes even if the argon pressure is two to four times higher than in the older arc tubes. The ignitor used in pulse start tubes shuts off once the lamp has started. The lower current reduces lumen depreciation.

The new ballasts are slightly more expensive, but a standard 400 watt lamp can use almost $1000 of electric power over the life of the lamp typically 20,000 hours. Up to 20% improvement in mean lumens per watt (LPW) is possible with the new systems. In continuous burn applications where lamps are turned off only once a week, a 400 watt pulse start system can provide up to 30,000 hours of lamp life.

Along with the new ballast with ignitor starting, improvements were made in lamp design. The newer formed body arc tubes are ellipsoidal (spherical) in shape with legs on either side. The tubular bodies are produced by pinching the two ends of a cylinder of quartz. There are only two electrodes, one at each end, eliminating the additional starter electrode at one end. Since there are only two electrodes, the formed body arc tubes need a high voltage starter pulse. An electronic high voltage pulse ignitor provides a pulse of 3000 to 4000 volts for starting.

The ignitor is usually included in the ballast circuit which allows these arc tubes to be filled with a higher pressure of argon gas which improves long term performance. Standard lamps can be used on ballasts with ignitor starting but they provide improvements only in lumen maintenance without the improved efficacy and color uniformity.

The combination of pulse start ignitor ballasts with formed body arc tube lamps provide higher lumens, better lumen maintenance, improved color uniformity, faster warm-up and hot restrike, and improved ballast efficiency.

Ballasts for pulse start systems include the older constant wattage autotransformer (CWA) ballast and the newer 277 volt reactor and regulated lag ballasts which use a high voltage ignitor to start the arc. These provide more reliable cold starting, hot restrike in 50% less time and reduced arc tube wall darkening. This results in improved lumen maintenance and longer lamp life.

Some reactor systems are almost twice as efficient as other ballasts and provide significant power cost savings as well. Reactor ballasts operate on 277 volts without a step-up transformer or autotransformer.

Ballasts must also provide regulation which protects the lamp performance from line voltage fluctuations. Reactor ballasts offer poor regulation. A 5% drop in line voltage can lead to a 10% drop in light output. Regulated lag ballasts with pulse start ignitors offer good regulation for metal halide lamps, although they are not as efficient as reactor ballasts.

The 277 volt reactor ballast has a lower current crest factor because there is very little distortion of the wave form. Pulse start reactor ballasts for use on 277 volt circuits reduce ballast losses by almost 50% by eliminating the need for voltage transformation. A 350 watt rated pulse start reactor ballast requires only 30 watts, while a standard 400 watt rated CWA ballast will use about 60 watts.

Replacing a standard 400 watt unit with a 350 watt pulse start unit on a reactor system will save 50 watts in the lamp and 30 watts in the ballast for a total reduction in power consumption of 80 watts or 80/460 = 17%. The lumen output, initially, will go up by 6% and the mean lumens, at the 8000 hour point, will go up by 25%. The average maintained efficacy of systems (AMES) is mean lumens divided by system watts. The current crest factor is 1.4 which allows improved lumen maintenance compared to the older technology 1.8 create factor.

Another improvement to the reactor ballast is called the controlled current reactor. It reduces the current during lamp starting and open circuit conditions. System energy savings can amount to over 25%.

The regulated lag ballast offers the greatest level of control over lamp wattage fluctuations as a result of input line voltage variation. This ballast is larger and heavier than either the 277 volt reactor or the CWA and is used in industrial applications where users are concerned with the effect of voltage fluctuations on lamp life.

OPEN FIXTURE LAMPS

Since metal halide lamps operate at high temperatures, standard metal halide lamps require enclosed fixtures due to the possibility of end-of-life arc tube ruptures which could release arc tube and bulb glass fragments from the fixture.

This danger is greater when lamps are burned 7 days a week, 24 hours a day. The metal halide lighting industry recommends that lamps be shut down once a week for at least 15 minutes. This helps to assure that end-of-life lamp failure occurs passively at low pressure or by simply failing to re-ignite. It should be done along with proper maintenance and group relamping before the end of rated lamp life.

It has not been proven that these steps eliminate nonpassive end-of-life rupture. So, new pulse start lamps specifically designed for open fixtures utilize a shrouded arc tube. This is sometimes called a double-containment feature to enclose shattered particles.

The design of the formed body arc tube allows a protective shroud inside the envelope. The heavy glass shroud around the arc tube provides containment of hot arc particles should a non-passive arc tube failure occur.

By eliminating the need to enclose the fixture, the open fixture rating allows more flexibility in design, lower fixture costs and improved maintained lumens. Fixture manufacturers can eliminate the glass lenses that reduce lighting efficiency due to light absorption and dirt accumulation and that creates potentially higher operating temperatures. Lamps with an open rated shrouded arc tube can be operated continuously with no weekly shutdown required.

ANSI has developed a test to determine if a lamp is safe for open fixture operation. In the ANSI test, lamps designated S are unshrouded and do not pass the test. Lamps designated O pass the ANSI test. The S lamps must be operated within 15° of vertical and must be turned off for 15 minutes every week.

Replacing the yellow light of high pressure sodium with the white light of pulse start metal halide lamps can provide a better quality, more natural light. Unlike the low color rendering index (CRI) of 22 provided by high pressure sodium lamps, pulse start lamps have a CRI range of 65-70. Pulse start systems with the formed body arc tubes are designed to the same light center lengths as the old standard lamps. This allows the use of the same fixture and reflector. Installing a new pulse start system requires a new pulse start ballast.

A pulse start ballast and formed body arc tube can provide 10% to 20% higher lumens than standard lamps with less lumen depreciation over time and no increase in energy usage. A lower wattage pulse start system can provide the same light output as the original system with significant energy savings.

In one retrofit project, a supermarket chain switched from standard 400 watt metal halide lamps operating on CWA ballasts to pulse start systems. The chain decided to maximize its energy savings by installing 320 watt pulse start systems. Combined with the 277 volt reactor ballast, the 320 watt lamp system provided a savings of 113 watts per fixture. The annual energy savings was over $10,000.

Tyson Foods is the world's largest poultry producer with 115 facilities in 17 states. Retrofit goals included a 30% reduction in lighting cost with an overall energy savings of 2% of their total electrical consumption. Recycling of lamps and ballasts would be done as required by state and federal regulations to meet Green Lights and EPA requirements. Other goals were to reduce product and maintenance cost, and use no up-front capital.

The existing system consisted of magnetic ballasts with T12 lamps and probe start metal halides. Most fixtures are mounted over process equipment. Because of routine exposure to high-pressure wash-downs, the lenses are susceptible to cracking and require frequent replacement along with sockets and wiring.

The installation of special fixtures to withstand the high water pressure solved many of the maintenance problems. The fixtures plug in to allow easy removal for repair. Special stainless conduit and rubber cords were used to protect the electrical system. New electronic ballasts and T8 lamps were installed. Each facility showed decreases in lighting repairs.

Manufacturer warranties were as follows:
- Fluorescent electronic ballasts - 5 years,
- Fluorescent T-8 Lamps - 2 years,
- HID Pulse Start Metal Halide Ballasts - 5 years,
- HID Pulse Start Metal Halide Lamps - 2 years.

SULFUR LAMPS AND LIGHT PIPES

Optical fibers composed of certain glasses or plastics can transport light efficiently, but are impractical for transporting large quantities of light. Large solid fibers or fiber bundles would be required. These bundles are heavy, difficult to install in many applications and very expensive. Fiber-optics were first produced and patented in the 1970s.

Prismatic materials in a hollow light guide can efficiently transport large quantities of light. The prism light guide was patented in 1981. The first guides were constructed as rigid rectangular acrylic pipes with molded prisms.

In 1993, 3M started making this macro-prism structure in the form of a continuous film. This is known as 3M Optical Lighting Film (OLF).

The losses due to absorption and transmission are small and the reflectance efficiency approaches 99%. Tube diameters up to 10 inches or larger are available.

A sulfur light bulb contains a small amount of sulfur and inert argon gas. When the sulfur is bombarded by focused microwave energy, it forms a plasma that glows very brightly, producing light containing all colors and closely matching that of the sun, but with very little heat or ultraviolet (UV) radiation.

The golf-ball sized bulb is rated at 1425 watts producing about 135,000 lumens with a color rendering index of 79. Since there are no filaments or other metal components, the bulb may never need replacement.

The sulfur lamp can be used with reflectors for high bay fixtures or with a hollow light guide commonly called a light pipe for illuminating large areas. The light emitted can be filtered, tinted, dimmed and reflected to meet precise lighting needs.

In 1970, Fusion Systems began developing an electrodeless, microwave-powered lamp for UV applications. The first commercial sulfur lamp was available in 1994. It was developed by Fusion Lighting and was supported by the U.S. Department of Energy (DOE) and NASA.

SULPHUR LAMP APPLICATION

Sulfur lamps are now being used in the U.S. and Europe for applications including airport tarmacs, aquariums, automobile assembly plants, cold storage facilities, gas stations, gymnasiums/sports facilities, highway signs, museums, postal sorting facilities and subway stations.

The first DOE-sponsored applications of sulfur lamps in the U.S. are at the Smithsonian National Air and Space Museum and on the outside of DOE's Forrestall Building, both in Washington, DC.

Hill Air Force Base wanted to upgrade the quality of lighting in the low bay and hangar areas to 70 foot-candles. The existing mercury vapor and metal halide light fixtures produced very low light levels for the tasks performed in the hangars and low bays. In the low bays, fixtures mounted at 26 feet produced only 40-45 foot-candles of light. In the high bays, 28-30 foot-candles were produced by fixtures mounted at 45 feet.

Hill AFB became the world's largest installation of sulfur lighting. It is used in aircraft hangars for F16 fighter and C130 cargo aircraft maintenance and overhaul. Light guides using Light Drive 1000 lamps were used in the lowbay area. Light sources were located over walkways and driveways so that no maintenance had to occur over aircraft.

The light guides use 10-inch-diameter tubes fabricated of multiple layers of plastic materials. Each 10-foot section of hollow light guide weighs approximately 30 pounds. Forty-four light pipes are installed, each being 105 feet long with a Fusion lamp module coupled to each pipe end.

In the large high bay hangars, a minimum mounting height of 45 feet is required to allow working on C-130 and F-16 aircraft. A high bay fixture of metal refractor globes lined with glass was used.

Hill AFB's sulfur fusion lighting retrofit reduced the number of fixtures in the low bay areas while increasing light levels to 70 footcandles using 88 sulfur lamps and 44 light pipes. In the hangars, the number of fixtures was maintained, but the light level was tripled. A total of 288 sulfur lamps was used.

Compared to the high-intensity discharge systems they replaced, the sulfur lamps produced lighting levels that were almost 50% higher in the low bay area and up to 160% higher in the high bay area. Comparable lighting levels with metal halide lamps would consume almost 20% more energy in the low bay area and about 40% more energy in the high bay area.

HIGH PRESSURE SODIUM (HPS)

These provide a higher efficacy than metal halide lamps and are popular for outdoor or industrial applications where good color rendition is not required. HPS is often used in parking lots. They produce a light golden color that allows some color rendition.

HPS should not be installed near metal halide lamps or fluorescent systems. These provide a white light which looks more normal than HPS. This separation has been used in indoor gymnasiums and industrial spaces.

LOW PRESSURE SODIUM

LPS systems have the highest efficacy of available HIDs, with values up to 180, but the monochromic light source provides a poor color rendition. With a low CCT, the color is a pumpkin orange and objects appear white, shades of gray or black.

The lighting quality can be improved by supplementing the LPS system with light sources having a greater CRI. LPS is popular because of its high efficacy. It provides up to 60% greater efficacy than HPS and has a life of about 18,000 hours.

Applications are limited to security or street lighting. The lamps are up to 3 feet long and are not good sources at high mounting heights. Cities, such as San Diego, CA, have installed LPS systems on streets and many parking lots in San Diego have also switched to LPS.

BALLASTS

Except for incandescent systems, lighting systems require a ballast. The ballast controls the voltage and current to the lamp and has a direct impact on the light output.

The ballast factor is the ratio of a lamp's light output to that achieved with a reference ballast. General-purpose fluorescent ballasts have a ballast factor that is less than one. It is about 9 for most electronic ballasts.

Higher ballast factors increase the light output and lower ballast factors reduce the light output. A ballast with a high ballast factor also consumes more energy than a general-purpose ballast.

Electronic ballasts for fluorescent lamps have been available since the early 1980s. In the early years of electronic ballasts, they suffered from reliability and harmonic distortion problems. Today, most electronic ballasts have a failure rate of less than 1%, and less harmonic current than magnetic ballasts. Electronic ballasts are typically 30% more energy efficient. They also produce less lamp flicker, ballast noise, and waste heat.

Magnetic ballasts include the standard core and coil, high-efficiency core and coil (energy-efficient ballasts) and the cathode cutout or hybrid type. Standard core and coil magnetic ballasts are core and coil transformers and are relatively inefficient at operating fluorescent lamps. These types of ballasts are no longer sold in the U.S. but they are still found in many facilities. The high-efficiency magnetic ballast can be used to replace the standard ballast and improve the unit efficiency by about 10%.

Cathode cutout or hybrid ballasts are high-efficiency core and coil ballasts with electronic circuitry to cut off power to the lamp cathodes after the lamps are operating. This provides an additional 2 watt savings per lamp.

In most fluorescent lighting applications, electronic ballasts can be used in place of conventional magnetic core and coil ballasts. Electronic ballasts improve fluorescent system efficacy by converting the standard 60-Hz input frequency to a higher operating frequency, usually 25,000 to 40,000-Hz. Lamps operating on these frequencies produce about the same amount of light, while consuming up to 30% less power than a standard magnetic ballast. Electronic ballasts have less audible noise, less weight, almost no lamp flicker and offer dimming capability.

The three major types of electronic ballasts are designed to operate with T12, T10 or T8 lamps. T10 ballasts consume about the same energy as T12 ballasts, but T10 systems provide more light output. Some T10 systems are more efficient than comparable T8 systems. Since T10s provide more light per lamp, fewer lamps need to be installed.

T8 ballasts are often more efficient, have less lumen depreciation and are more flexible. T8 ballasts can operate one, two, three or four lamps. Most T12 and T10 ballasts can only operate one, two or three lamps. One T8 ballast can replace two T12 ballasts in a 4 lamp fixture. T8 systems and parts are more likely to be available at lower costs.

Some electronic ballasts are parallel-wired, so that when one lamp burns out, the remaining lamps in the fixture will continue to operate. Typically, magnetic ballasts are series-wired so when one component fails, all lamps in the fixture are off. Before the lamp can be replaced, it must be determined which lamp failed. The electronic-ballasted system will reduce relamping time since you can see which lamp failed. Parallel-wired ballasts also allow reducing the lamps per fixture if an area is over-illuminated.

HID BALLASTS

High intensity discharge lamps also require ballasts to operate. HID ballasts are available in dimmable, bilevel output and instant restrike versions.

Capacitive-switched dimming can be installed as a retrofit to existing luminaires or as a direct luminaire replacement. Capacitive switching HID luminaires have bilevel capabilities and are designed to provide either full or partial light output based on inputs from occupancy sensors, manual switches or scheduling systems. Capacitive switching HID upgrades can be less expensive than using a panel type variable voltage control for dimming especially in circuits with few luminaires.

Typical applications of capacitive switching include athletic facilities, occupancy-sensed dimming in parking lots and warehouse aisles. Transmitters can also be used with other control devices such as timers and photo sensors to control the bilevel luminaires.

After detecting motion, the occupancy sensor signals the bilevel HID ballasts. The system then brings the light level up from a standby reduced level to about 80% of full output, followed by a warm-up time to 100% of full light output if desired.

The standby lumens are about 15-40% of full output and the standby wattage is 30-60% of full wattage. When the space is unoccupied and the system is dimmed, you can have energy savings of 40-70%. However, lamp manufacturers do not recommend dimming below 50% of the rated input power.

LUMINAIRES

The luminaire or light fixture contains the lamps, ballasts, reflectors, lenses, louvers and housing. Their main function is to focus or spread the light from the lamps. Without luminaires, lighting systems may have bright spots and produce glare.

Luminaires block or reflect some of the light exiting the lamp. The efficiency of a luminaire is the percentage of lamp lumens that leaves the fixture in the desired direction.

Efficiency depends on the luminaire and lamp configuration. Using four T8 lamps in a luminaire can be more efficient than four T12 lamps because the T8 lamps are thinner and allow more light to be released from between the lamps.

The Coefficient of Utilization (CU) is the percent of lumens produced that actually reach the work plane. The CU depends on the luminaire efficiency, mounting height, and reflectance of walls and ceilings.

Using reflectors on most luminaires can improve their efficiency. Reflectors perform better when there are less lamps or smaller lamps in the luminaire. A common luminaire upgrade is to install reflectors and remove some of the lamps in a luminaire. The luminaire efficiency is improved, but the light output from each luminaire may be reduced.

Reflectors can redistribute light, usually more light is reflected downward, which can produce bright and dark spots in the space. The performance from reflectors drops as they get dirty. As dirt accumulates on the surfaces, the light levels will drop.

A variety of reflector materials are used including reflective white paint, silver film laminate, and anodized aluminum. Silver film laminate has the highest reflectance, but is less durable.

Besides installing reflectors in luminaires, light levels can be increased by improving the reflectiveness of the room's walls, floors and ceilings. Covering a dark wall with white paint reflects more light back into the workspace.

Most indoor luminaires use either a lens or louver to prevent occupants from directly viewing the lamps. Lenses and louvers are designed to shield the viewer from direct beams of light.

Lenses are usually sheets of hard plastic, either clear or milk white, that are located on the bottom of a luminaire. Clear, prismatic lenses are 65% efficient since they allow most of the light to pass through, and less light to be trapped in the fixture.

Milk white lenses act as diffusers and are less efficient, since they trap more light in the fixture. Although diffusers are used in many office environments, they have a lower efficiency of about 50%.

Louvers provide better glare control compared to most lenses. A louver with a grid of small plastic cells shields some of the horizontal light exiting the luminaire and reduces the glare. Generally, the smaller the cell, the less the efficiency. The efficiency is about 40-45% for the smaller cells. Deep-cell parabolic louvers provide a better efficiency (75-95%) but they require deep luminaires.

LIGHT DISTRIBUTION AND MOUNTING HEIGHT

Fixture mounting height and light distribution are interactive. HID systems are used for high mounting heights since the lamps act as a point source and reflectors can direct light downward with a high level of control. Fluorescent tubes are diffuse sources, with less ability to control light at high mounting heights. Fluorescent systems are better for low mounting heights and/or areas that require diffuse light with minimal shadows.

High-bay HID luminaires are used for mounting heights greater than 20 feet high. These luminaires usually have reflectors and focus most of the light downward. Low-bay luminaires are used for mounting heights less than 20 feet and use lenses to direct more light horizontally. The following minimum heights should be used:

*   1000 W High pressure sodium ...... 26 feet
*   400 W High pressure sodium .... 18 feet
*   200 W High pressure sodium .... 15 feet
*   1000 W Metal halide .... 20 feet
*   400 W Metal halide .... 16 feet

HIDs are potential sources of direct glare since they produce a large mount of light from a physically small point source. The problem of excessive glare can be minimized by mounting the fixtures at sufficient heights.

In a direct lighting system, 90-100% of the light is directed downward. In an indirect lighting system 90-100% of the light is directed to the ceilings and upper walls. A semi-indirect system would distribute 60-90% downward, with the remainder upward.

EXIT SIGNS

Recent advances in exit sign systems provide new opportunities to reduce energy and maintenance costs. Since emergency exit signs operate 24 hours a day, energy savings can quickly recover the retrofit costs.

Many retrofit kits have adapters that screw into the existing incandescent sockets. Installation takes only about 15 minutes per sign. However, discolored or damaged signs should be replaced in order to maintain the illuminance required by fire codes.

Upgrade technologies for exit signs include

- Compact Fluorescent Lamps (CFLs),
- Incandescent assemblies,
- Light Emitting Diodes (LEDs),
- Electroluminescent panels, and
- Self luminous tubes.

Replacing incandescent sources with compact fluorescent lamps was the first of the exit sign upgrades. Most CFL kits require rewiring. CFLs consume about 10 watts and have a life close to 2 years.

More advanced upgrades provide reduced maintenance costs, greater efficacy and even installation in low sub-zero temperature environments. LED upgrades are the most cost-effective since they consume less than 4-W and have a life of more than 25 years, almost eliminating maintenance. An LED retrofit involves a pair of LED strips on the inside of the exit sign enclosure that provide a soft, glowing sign.

Another low-maintenance upgrade is a string of incandescent assemblies. These are low-voltage luminous strings or ropes that consume about 8 watts and use existing sockets like LED retrofit kits. Incandescent assemblies can have bright spots that are visible through the transparent exit sign. They have a life of over 3 years.

Electroluminescent panels consume less than one watt, but the light output depreciates over time. The panels have a life of more than 8 years. The self-luminous sources are the most energy-efficient since they require no electricity. But, the spent tritium tubes which are used to illuminate the unit must be disposed of as radioactive waste, which increases the overall costs. The luminous tubes have a life of 10 to 20 years.

WORKSPACE-SPECIFIC LIGHTING

Illumination requirements for reading and writing are considerably higher than illumination requirements for general circulation areas. Traditional office lighting systems usually furnish a uniform light level throughout the area. Desk lighting is often limited to module lighting with small fluorescent lamps or other types of personal desk lights that can be moved to where the visual tasks are located. Although this type of task lighting can be adequate. Shadows can be prevalent, and the worker must continually adjust between a bright desktop area and darker areas away from the desktop.

A better solution is workspace-specific, direct-indirect lighting. A suspended luminaire over each workspace can furnish both task and ambient lighting. Reading and writing surfaces are illuminated by the downlighting component of the luminaire. Uniform ambient lighting is provided by the uplighting as it reflects off the ceiling and brightens the walls and ceiling.

Workspace-specific lighting provides task illumination with minimum energy use. A 10x12' area can be illuminated with two T8 2-lamp or 3-lamp fluorescent fixtures. The power density will be in the range of 1.0 to 1.5 watts per square foot. In the workspace-specific approach, one 3-lamp direct-indirect luminaire is used with up to a 50% reduction in power density or about 0.75 watts per square foot. In open areas, the power density can be as low as 0.6 watts per square foot.

This means workspace lighting can require fewer than half of the fixtures to light the same area. The energy-efficiency is due in part to the efficiency of direct-indirect luminaires in delivering light to the work surfaces.

In the proximity of the reading and writing visual tasks—desk areas—the illumination is generally in the range of 40-60 foot-candles. The illumination gradually drops off beyond the visual task area to an ambient level of 20-30 foot-candles where circulation and short-duration visual tasks may be performed.

Since each workstation receives the same fixture placement, each worker receives the same quality and quantity of light. This solves a common problem in traditional open-office plants where some partitioned workstations are in the shadows. The direct-indirect lighting distribution also brightens the ceiling and upper walls, reducing the cave effect typically connected with the sharp cutoff of parabolic louvers.

LIGHTING CONTROLS

Lighting controls provide the ability for systems to be turned on and off either manually or automatically. Control upgrades range from simple manual switches in the proper locations to sets of occupancy sensors and timers.

The disadvantage with manual switches is that people often forget to turn them off. If switches are far from room exits or are difficult to find, occupants are more likely to leave lights on when exiting a room. If switches are located in the right locations, with multiple points of control for a single circuit, occupants find it easier to turn systems off.

Another opportunity for upgrading occurs when the existing lighting system was designed to be controlled from one switch, when only some circuits are often needed. Since the lights are all controlled from one switch, every time the area is used all the lights are turned on. By dividing the circuits and installing more switches substantial energy savings are possible. If it is not too difficult to re-circuit the existing lighting system, additional switches can be added to optimize the lighting controls.

Time clocks can be used to control lights when their operation is based on a fixed operating schedule. Time clocks are available in electronic or mechanical units. Regular check-ups are needed to ensure that the time clock is controlling the system properly. After a power loss, electronic timers without battery backups can be off schedule and cycle on and off at the wrong times.

For most outdoor lighting applications, photocell controls which turn lights on when it gets dark, and off when sufficient daylight is available offer a low-maintenance alternative to time clocks. Unlike time clocks, photocells are seasonally self-adjusting and automatically switch on when light levels are low, such as during rainy days.

Photocells work well in almost any climate, but they need to be clean. Dust can accumulate on the photocell aperture causing the light to stay on in a cloudy day.

A photocell is inexpensive and can be installed on each luminaire, or can be installed to control numerous luminaires on one circuit. Photocells can also be effectively used indoors, if daylight is available through skylights.

WORKSPACE CONTROLS

Direct/indirect light fixtures also allow new opportunities to maximize the energy savings with automatic lighting controls. Workspace lighting allows convenient personal dimming controls with dimmable electronic ballasts.

Occupancy sensing can be used with automatic dimming or switching based on occupant motion within the workspace. Daylight dimming and control is a type of automatic control in response to changing daylight conditions.

Since the light fixture is directly over the task area, the occupancy sensor can be placed in the fixture itself. The distance between the sensor and occupant movement is minimized.

A concern in open offices is reducing the sudden changes in room brightness caused by individual fixtures switching on or off. Occupancy sensors can be set to control the downlight, task lighting, while keeping the uplights, ambient lighting, on at a constant level. To minimize distractions to neighbors in an open office, the downlighting can be gradually dimmed after the space becomes unoccupied. A time delay could be used to turn the downlighting off. This would allow a constant ceiling brightness with progressive changes in downlighting.

Occupancy sensors save energy by turning off lights in spaces that are unoccupied. The savings for various areas are shown below:
• office areas 20-50%,
• meeting rooms 45-65%,
• corridors 30-40% and
• warehouse and storage areas 50-75%.

When the sensor detects motion, it activates a control device to turn on the lighting system. If no motion is detected in a specified period, the lights are turned off until motion is sensed again. Both the ability to detect motion and the time delay difference between no motion and when the lights go off are adjustable.

The basic types are Ultrasonic (US) and Passive Infrared (PIR). Dual-technology sensors, that have both ultrasonic and passive infrared detectors, are also available.

Sensors are available as wall-switch units or remote units for ceiling mounting or outdoor applications. A low-voltage connection is used for the sensor to the control module. In wall-switch sensors, the sensor and control module are one unit. Multiple sensors and lighting circuits can be wired to one control module. Wall-switch sensors can replace existing manual switches.

Wall-switch sensors need an unobstructed path for coverage of the area to be controlled. Ceiling-mounted units can be used in corridors or open office areas with partitions. There are also sensors for outdoor use.

Ultrasonic sensors transmit and receive high-frequency sound waves that are above the range of human hearing. Motion in the room distorts the sound waves. The sensor detects this distortion and signals the lights to turn on. If no motion is detected over a specified time, the sensor sends a signal to turn the lights off. Ultrasonic sensors need enclosed spaces for sound waves to echo the reflections. They can only be used indoors and perform better if the room surfaces are hard and the sound wave absorption is minimal.

Passive infrared sensors detect differences in the infrared energy in the room. As a person moves, the sensor detects a heat source moving. These sensors need an unobstructed view, and as the distance from the sensor increases, larger motions are needed to trigger the sensor.

Dual-technology sensors combine both US and PIR sensing technologies. They can improve sensor reliability and minimize false switching.

COMPUTER DIMMING CONTROLS

Most workers prefer to adjust light levels down from typical office levels. Intelligent lighting control systems are now available that allow workstation computers to control light fixtures.

An on-screen control panel with click-and-drag dimming slider controls will lower workstation illuminance. This results in less complaints and better working conditions for maximum productivity.

The use of local area networks (LANs) for lighting control allows centralized lighting control from network computers. The primary modes of central lighting control are load shedding and scheduling. Load shedding allows customers to take advantage of low rate structures. The technique is to turn off nonessential electric loads during high-cost electricity use periods, but lighting is rarely considered non-essential. By gradually dimming the lighting system, typically 20-30% over 15-20 minutes, the on-peak lighting power costs can be reduced by up to 30% without distracting workers. The human eye adapts to changes in brightness much faster than the load-shedding dimming rate and the change in brightness will not be evident.

Reduced light levels should only be a temporary measure for minimizing the peak electricity demand. Full illumination levels should be restored at the same gradual rate after the peak demand period has passed. This type of load shedding takes advantage of time-of-use or real-time pricing rates.

Light scheduling is used to ensure that manually controlled lighting systems are off at the end of the day. Occupancy sensors can be used to control the downlighting and a scheduling control used to turn off the uplighting at the end of the day and turn it back on for the next workday.

There must be some way to temporarily override the schedule for special events. An occupant may need to work beyond the scheduled lights out time. The occupant should be able to override the lights out command in their area.

Light scheduling controls can also be used for scheduled dimming. The light may need to be scheduled to stay on during cleaning times. This can be done at a reduced light level. Load shedding may also need to be scheduled at specific times.

Other LAN-based lighting control functions include global controls commissioning and energy use monitoring. These can be implemented from a central computer, instead of using individual sensors in the workspace and installing individual energy monitoring devices throughout the building.

Occupancy sensors can affect lamp life since they may cause rapid on/off switching. This reduces the life of some fluorescent lamps which lose 25% of their life if turned off and on every three hours.

Although occupancy sensors may cause lamp life to be reduced, the actual on-time also decreases and in many cases, the time until relamp will not change. However, the use of occupancy sensors should be evaluated if the lights are to be switched on and off frequently. The longer the lights are left off, the longer the lamps will last.

Occupancy sensors are not used with HIDs, but some HID ballasts offer bilevel control to dim and re-light lamps quickly. This is done with capacitive switching HID luminaires.

Occupancy sensors save the most energy when used in rooms that are not used for long periods of time. The advantage of occupancy sensing is achieved because private offices and workstations can be unoccupied for over 40% of the time that lights are normally on.

DAYLIGHT DIMMING

Daylight dimming can provide a nearly constant level of work surface brightness as daylight contributes variable amounts of illumination. A light sensor detects the luminance or brightness of the work surface and controls the light output to maintain the illumination level. The rate of dimming should proceed at a slow rate of 30-60 seconds from maximum to minimum light output.

Daylight dimming can provide an energy savings of 30% in window-adjacent locations. The energy savings typically occur during periods of peak electricity demand, resulting in reduced utility demand charges.

EELA

The Energy Efficient Lighting Association (EELA) promotes lighting retrofits with the latest technology. It started in 1997 and is based in Princeton Junction, NJ. The EELA sponsors lighting conferences and is supported by lighting service companies.

In one study for a large hospital, the EELA recommended a lighting retrofit with electronic ballasts and energy-efficient lamps. A local utility offered rebates for part of the costs. In a few years, the hospital found that it had saved several hundred thousand dollars in energy costs.

Companies such as Prescolite are incorporating feedback from installers, end-users, and design professionals in energy-efficient light product design. There are now energy-saving universal voltage ballasts that accept multiple wattages and lamp types. Other developments include advances in dimming ballasts for compact fluorescent lamps with direct digital dimming and end-of-life circuitry. Microprocessor control is used in the ballasts for reducing energy costs and providing flexibility for end-users. Updated ballast technology can be used with group relamping.

Group relamping is the opposite of spot relamping and it saves on maintenance, as well as utility costs. Many facilities use piecemeal relamping.

Retrofitting provides an average energy savings of 50 cents for each square foot of building space. The average payback is 2-1/2 years. Partnering with energy service companies can also help with energy audits and funding lighting upgrades.

RECENT LED LIGHTING ADVANCES

The incandescent light bulb has been a reliable source of artificial illumination since Thomas Edison completed the design in 1878. Just as vacuum tubes were replaced by integrated circuits in the late 1950s, the light bulb may soon be replaced by a new generation of solid-state, light-emitting electronic devices. Light-emitting diodes (LEDs) that provide red light are already replacing the bulbs in traffic lights and automobile taillights. LEDs and semiconductor lasers can also emit blue, green, and yellow light which are used in new types of flat-panel displays for applications such as exit signs.

LEDs convert electricity to colored light more efficiently than incandescent bulbs, for red light, their efficiency is 10 times greater. They are rugged and compact. Some types can last for 100,000 hours, which is about a decade of regular use. In contrast, the average incandescent bulb lasts about 1,000 hours.

The intensity and colors of LED light have improved so that the diodes are now used for large displays such as the eight-story-tall Nasdaq billboard in New York's Times Square. It uses over 18 million LEDs and covers 10,736 square feet.

LEDs are already replacing light bulbs in automotive applications. In Europe 60 to 70% of the cars produced use LEDs for their high-mount brake lights. LEDs are also being used for tail light and turn signals, as well as for side markers for trucks and buses. LEDs will dominate the red and amber lighting on the exterior of vehicles.

Larger and brighter LEDs are being used in traffic lights. Traditionally, traffic signals and other colored lamps use incandescent bulbs, which are then covered with a filter to produce the appropriate color. Filtering is cheap but it is inefficient. A red filter blocks about 80% of light as the amount of light that emerges drops from about 17 lumens per watt of power to about three to five lumens per watt.

LEDs consume only 10 to 25 watts, compared with 50 to 150 watts used by an incandescent bulb of similar brightness. This energy savings pays for the higher cost of an LED in as little as one year. This must be considered along with the reduced maintenance costs of LEDs.

Interior designers having been using LEDs for several years, when high-brightness models of various colors appeared. Since each LED gives off a distinct hue, users have control of nearly the full spectrum.

By combining different colored LEDs together in an array, the user can adjust their combined light. The white light from an array of red, green and blue LEDs can be made cooler by increasing the light from the red LEDs and reducing the light from the blue ones.

This flexibility offers new ways of using light. Instead of changing wallpaper or paint, the color of a room can be changed by adjusting the ratio of wavelengths in the emitted light. The wavelengths of light determine the colors. The Metropolitan Museum of Art in New York City used this type of LED lighting to illuminate its display of the Beatles' Sgt. Pepper's costumes in 1999. The light is also cool and will not damage the fabric. Photographers would also have total control of their light sources without the need for filters or gels.

Until recently, solid-state devices could not provide the customary white light that illuminates homes and workplaces. One way to generate white light from LEDs is to combine the output of LEDs at the red, green and blue wavelengths. It is difficult to mix the color of LEDs efficiently with good uniformity and control.

White light can also be obtained by using an LED photon to excite a phosphor. A yellow phosphor can be placed around a blue LED. When the energy of the LED strikes the phosphor, it is excited and gives off yellow light, which mixes with the blue light from the LED to give white light.

Another way is to use an ultraviolet LED to excite a mixture of red, green and blue phosphors to give white light. This process is similar to that used in fluorescent tubes. It is simpler than mixing colors but is less efficient since energy is lost in converting the ultraviolet or blue light into lower-energy or light toward the red end of the spectrum. Light is also lost due to scattering and absorption in the phosphor packaging.

A type of microlaser known as VCSELS (vertical cavity surface-emitting laser) is used to produce ultraviolet radiation. Ultraviolet light can cause the phosphors on the inside of gas-filled fluorescent lights to glow white. The phosphorescent coating can turn arrays of tiny lasers into practical lamps. Many groups are working to create such lasers in the UV range. Arrays of tiny lasers have been developed by researchers at Brown University and Sandia National Laboratories.

VCSELS create a monochromatic and directional beam although there are other approaches. Blue LEDs can be combined with phosphors to create white light and blue LEDs are already in production. In the U.S., GelCore which is a joint-venture between GE Lighting, Emcore CreeLighting are working on this approach. This light tends to be harsh and cold, since it is heavy in the blue spectrum. White light also can be produced by arrays that combine the spectral ranges of red, blue, and green LEDs, but they are more costly to fabricate.

About $15 billion worth of conventional lamps are sold each year worldwide, but solid-state lights would be superior. They can last up to 10 times longer than fluorescent tubes and can be far less fragile.

Low-power white LEDs with an efficiency slightly better than incandescent bulbs are available commercially, but high-power devices suitable for illumination are still too expensive.

Instead of fragile, hot, gas-filled glass bulbs that burn out relatively quickly and waste most of their energy in the form of heat, LEDs would provide long-lasting, solid-state interior light. In automobiles, LEDs could last the lifetime of the car.

Lighting represents about 30% of the U.S. electrical use, and even the best standard illumination systems convert only about 25% of electricity into light. If white LEDs could be made to match the efficiency of today's red LEDs, they could reduce energy needs.

There is about a $12-billion worldwide market for illumination lighting. The big three of lighting: Philips, Osram Sylvania and General Electric are working hard on LED research and development and newer companies such as Lumildes which is a joint venture between Philips and Agilent Technologies are appearing.

Low-power white LEDs are already being used in cell phones and pedestrian walk signals. Second-generation, higher-power LEDs for accent lighting are also becoming available. They may also change the way spacelighting is used. Flat arrays could be mounted in various patterns, on floors, walls, ceilings, or even furniture.

The biggest advantage may be in saved energy. Researchers at Sandia and Hewlett-Packard predict that solid-state lighting could reduce energy costs by $100 billion annually by 2025. The demand for electricity could be reduced by 120 gigawatts per year which is about 15% of current U.S. generating capacity. Solid-state illumination for civilian and military applications is being pursued in Japan, Taiwan, South Korea, and Europe.

LEDs also offer some possible advances for medical treatment. The LEDs' cool temperatures, precise wavelength control and broad-beam characteristics are allowing cancer researchers to use photodynamic therapy for tumor treatment.

In this type of therapy, patients receive light-sensitive drugs that are preferentially absorbed by tumor cells. When light of the appropriate wavelength strikes these chemicals, they become excited and destroy the cells. An array of LEDs produces an even sheet of light that stimulates light-sensitive drugs without the fear of burning the patient's skin.

Today's light-emitting diodes use a reflector, which holds the semiconductor chip (Figure 7-2). In LEDs, the crystal is not silicon but a mixture of group III and group V elements. By controlling the concentration of aluminum, gallium, indium and phosphorus and incorporating dopants, such as tellurium and magnesium, the formation of the n- and the p- layers takes place. The layer that has an excess of electrons is called n-type since it has a negative charge. Another layer uses a material that has fewer electrons which results in an excess of positively charged particles known as holes. This material is p-type since it has net positive charge. At the junction of the n and p layers an active layer exists where the light is emitted.

Applying a voltage across the junction forces electrons and holes into the active layer, where they combine and emit photons, the basic units of light. The atomic structures of the active layer and adjoining materials on each side determine the number of photons produced and their wavelengths.

Early LEDs were made in the 1960s with a combination of gallium, arsenic and phosphorus to yield red light. In these devices electrons combined with holes relatively inefficiently. Every 1,000 electrons produced one red photon. These LEDs generated less than one tenth the amount of light found in a comparably powered, red-filtered incandescent bulb.

By 1999 Hewlett-Packard was building LEDs that transformed more than 55% of the incoming electrons into photons at the red wavelength. Improvements in material quality and the development of other substances allows the more efficient transformation of electrons and holes into photons.

It was found that the materials do not have to be homogeneous. Each layer can have a different chemical makeup, but when placed

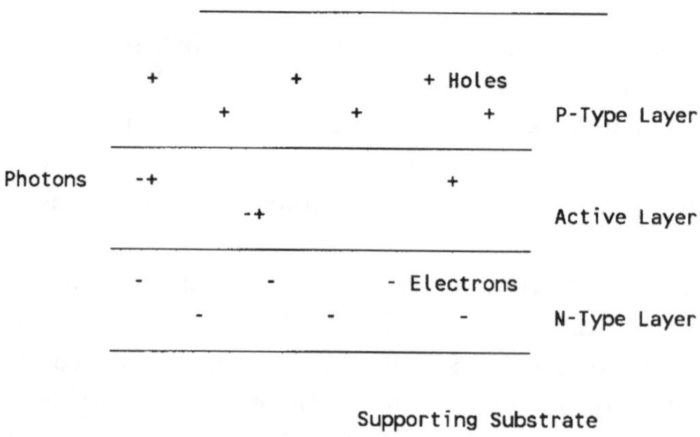

```
        _____

             +          +         + Holes
                 +          +          +      P-Type Layer
                _____
Photons      -+                        +
                   -+                          Active Layer
                _____
             -          -         - Electrons
                 -          -          -       N-Type Layer
        _____


                 Supporting Substrate
```

Figure 7-2. Structure of light-emitting diode (LED)

next to the active layer they tend to confine the electrons and holes which increases the number of electrons that combine with a hole to produce light.

Another key to LED improvement lies in manufacturing techniques that create smooth crystals instead of those with defects. The atomic lattices of the p and n materials must match with those of the supporting substrate and active layer. One manufacturing method is vapor phase deposition, where hot gases flow over a substrate to create a thin film. This technique was improved by directing cool gases over a hot substrate, a process which was first used to produce semiconductor lasers. This process allows the growth of a wider variety of materials and is now used to make high-quality LEDs.

In the mid-1990s it was found that brightness could be improved by reshaping the chip. The original gallium arsenide wafer on which the active layer was grown is replaced with a transparent gallium phosphide wafer and the LED takes the shape of an inverted pyramid. This shape decreases the number of internal reflections and increases the amount of light from the chip.

Commercial white LEDs now cost about 50 cents per lumen, compared with a fraction of a penny per lumen for a typical incandescent bulb. As energy prices rise, LEDs should become more attractive.

MAINTENANCE OF LIGHTS

Dust and dirt can decrease lighting levels by up to 30% and shorten bulb life by increasing the heat of the fixture. Clean luminaires and bulbs regularly.

Many fixtures are designed for bulbs up to a certain wattage. Beyond this they may overheat and fail. Fluorescent lamps that begin to flicker should be replaced. The ballast continues to draw energy even when the bulb is not working properly.

References

Craford, M. George, Nike Holonyak, Jr. and Frederick A. Kisk, Jr., "In Pursuit of the Ultimate Lamp," *Scientific American*, Vol. 284, February 2001.

Rosenberg, Paul, *The Alternative Energy Handbook*, Lilburn, GA: Fairmont Press, Inc., 1993.

Scientific Staff of the Massachusetts Audubon Society, *The Energy Saver's Handbook*, Emmaus, PA: Rodale Press, 1982.

Sioros, Donna, Editor, *Building for the 21st Century: Energy and the Environment*, Lilburn GA: The Fairmont Press, Inc., 2000.

Turner, Wayne C., *Energy Management Handbook*, The Fairmont Press, Inc.: Lilburn, GA, 1997.

"Will Light Bulbs Go the Way of the Victrola?," *Business Week*, Vol. 1 No. 1, January 22, 2001.

CHAPTER 8

COMPUTER TECHNOLOGY

ENERGY CONTROL TRENDS

The growth of energy demand lies in the growth of population and per capita energy use. These are increasing faster than the mature markets in the developed world. Infrastructure of all types is much less energy efficient and is generally in need of replacement and expansion. Even though growth and technology transfer are projected for the long-term, market conditions are no longer controlled to provide a stable environment in the short-term.

Development during the middle of the 20th century focused on creating industrial infrastructure for national monopolies. In the 21st century, development requires new technology in the energy sector. This includes cleaner fuels and renewable sources, process alterations for high efficiency and low emissions, improved process monitoring and control, new transport systems and process and product design for low energy/ material use and reusability.

A few of these changes apply to central power stations but the emphasis is on end-user characteristics. End-users become a more active part of the energy sector rather than passive consumers. There will be a number of participants making independent decisions and the markets will be influenced, but not directed, by public sector incentives and rules.

Deregulation of the utility industry has led to numerous mergers and acquisitions. A group of global corporations is emerging, geared to competition and replacing the structure of local regulated monopolies. The debt crises of the 1980s along with the free market philosophy of the 1990s has led to restructuring of the public sector.

The independent power producers (IPP's) became a serious part of the utility industry during the 1980s. They offered relatively small, modern plants and altered the basic industry concepts of how power resources could be procured.

Even more radical was an expanded concept of demand-side management. Under regulatory mandates and incentives, utilities procured blocks of guaranteed energy conservation. This demonstrated that end-users and third-part service providers could be structured to provide improved end-use power efficiency. Major consumers find themselves no longer captive and the trend is for large users to analyze their energy use and shop among various suppliers. To serve these customers, specialist brokers and marketers have emerged. This has created new packaged products for rates and services.

ENERGY ASSESSMENT

Since 1978, the U.S. Department of Energy has sponsored the Industrial Assessment Center program. This program provides no-cost industrial assessments to small and medium-sized manufacturers around the United States. The program utilizes 30 universities which each perform 30 assessments annually at firms within 150 miles of their respective campuses. It has been successful in the training of energy efficiency and waste minimization techniques. The program has also been successful in helping manufacturers. The average annual savings is approximately $40,000.

For a manufacturer to qualify for an assessment, it must meet three of the four following criteria: less than 500 employees, less than 75 million dollars in gross sales, no in-house energy expertise and total energy bills between $100,000 and $1.75 million. They must also have a standard industrial classification (SIC) code of 20-39.

Manufacturers that are too small to qualify for the program may qualify for energy assistance through a program like the Maine Department of Economic and Community Development's Energy Conservation Program (DECD). Technical assistance is furnished by this type of state program.

The U.S. Department of Energy has developed publications in conjunction with Rutgers University for energy conservation. They have also developed an assessment workbook for manufacturers in energy efficiency.

CONNECTIVITY AND NETWORKS

Connectivity is required for the mixture of computers, telecommunications and the Internet essential for up-to-date business operations. Markets are now created and commerce conducted in virtual space.

There has been a major shift in the implementation and utilization of energy control systems. This shift has gone from discrete sensors and analog devices to more intelligent, processor-based networks of sensors and devices that return both data values and diagnostic information.

These intelligent building automation solutions are based on device bus networks. Open PC technologies such as Ethernet are coupled with device bus networks. This allows desktop, industrial, and laptop computers to be powerful connection tools.

In today's energy management climate, increasing costs and the need for improved efficiency of building systems require a new look at building automation systems. Many industries must also comply with new environmental and safety requirements.

In such a business environment, less of the required data are control specific. More information of other types is needed to maintain, diagnose and modify the control system. Intelligent instrumentation allows better performing devices to provide functions such as advanced diagnostics.

Intelligent devices also provide the flexibility to apply control centrally or at local processing points for improved performance and reliability.

BUILDING MANAGEMENT

Building automation, wireless technology and the Internet are changing the business of managing buildings and their energy use. Tenants may want to call the building on a Sunday night and program it to be running when they arrive for some needed after-hours use. They may also want an integrated security system that makes them feel safe without invading their privacy. This increased functionality is possible with a building communications technology like BACnet or LonWorks. This type of bus technology allows products of various building systems to communicate.

In one school district's main energy management system, a BACnet gateway allows the district's energy management system to transmit a setpoint to one or more of the rooftop unit controllers. In another project, BACnet is the key to communication between an ice rink's building automation system (BAS) and the controls for the chillers that make the ice in the rink.

The city of Memphis, Tennessee, has installed new control systems as an upgrade to its Fairgrounds Complex. The retrofitted controls are BACnet-compatible, and will be networked into a centrally monitored system via the city's existing municipal system.

Another project is a centralized network of BACnet-based building automation systems in Tucson, Arizona for about 25 municipal facilities. The new system replaces a group of 5- to 10-year-old direct digital control systems. The city was concerned about the reliability of their existing control systems and were expecting major system failures.

The project uses BACnet at the head end and the main distributed processors, connecting all of the buildings into a single centralized system. This allows the flexibility to allow true competitive bidding for control system additions in the future.

The U.S. General Services Administration (GSA) is using BACnet controls from Alerton Technologies and the Trane Company at its 450 Golden Gate Building, located in San Francisco. This is a large-scale implementation of the BACnet open communications protocol which replaces an older pneumatic control system.

The building houses several government departments and agencies, including GSA and federal courts. It has 22 floors with 1.4 million square feet and is one of the largest buildings in San Francisco. Each floor is larger than a football field. The building-wide energy management control system cost $3.5 million and should save over $500,000 in energy costs per year.

The Trane Tracer summit system is used with a BACtalk for Windows NT server. The Windows NT workstations use a graphics package to allow the operator to view and change the system information. An Ethernet 10Base2 LAN connects the system workstations and network controllers. A network hub is used on each floor with remote terminals connecting to the network through the Ethernet LAN.

The retrofit involved over 800 dual-duct and 60 single-duct VAV terminal units with BACtalk controllers. Each controller is programmable and communicates on the BACnet MS/TP LAN. The pneumatic operators have been replaced with electronic actuators and the pneumatic thermostats replaced with intelligent digital-display wall sensors. The eight main dual-duct air handling units have been retrofitted with programmable controllers.

Other HVAC systems like the one in the Knickerbocker Hotel in Chicago have reduced operating costs by automating the building's mechanical systems with a new LonWorks-based building control system. Sharing a single, common twisted-pair (78-kbits/s) wire backbone among elements, the system devices work as intelligent zone controls.

Instead of individual wires, the functions are handled by messages over a network. The plug-and-play device characteristics allow easy connection of HVAC components. The building automation involves multi-zone air handling devices, refrigeration equipment, and the chilled and hot water systems. The volume of air flowing into the lobby area is controlled by variable frequency drives. The control of dampers and zones is done with LonWorks-based intelligent actuators. In areas such as the ballroom, the lobby, and the restaurant, intelligent space sensors are used to transmit network messages back to the actuators that control the zone dampers.

In the air handling system, the separation of the hot and cold deck dampers caused a synchronization problem which was solved by retiming the neuron chips in the actuators. This was adjusted so that both dampers would always operate in the correct positions. The operation of the heating and cooling valves are also under LonWorks control.

The system operates in a Windows environment and allows the monitoring of control system data, fan temperatures, equipment switching and adjustment of operating parameters. Dedicated channels on the CATV backbone used are available for guest services such as monitoring the mini-bar, controlling the heating, and providing real-time security through a door locking system.

ENERGY SERVICES AND BUNDLING

Utilities have traditionally been producers and distributors of an energy commodity under regulated monopoly conditions. They are now in the process of re-envisioning their business. The trend is new services in response to deregulation in home markets. This new energy services model will become a major utility practice as competitive markets evolve.

In the new services model knowledge becomes crucial as marketing and sales packages are tailored for market segments and even specific end-users. In order to obtain long-term customer commitments under competitive conditions, bundling energy services will become significant. Methods of energy performance contracting will be integrated into utility marketing as a way to add customer value. This added value comes from expertise in managing energy and related resources.

The utility may re-direct its resources to offer new kinds of products, modeled after products from the financial services industry. These may include strategic planning for energy procurement, energy risk management and annuity plans, information and expertise sourcing and best practice reference, energy cost reporting and analysis. Other offerings might also include a software interface to other business systems and linking customers for industrial ecology developments.

Partnerships may develop between plant operations and utility operations to include on-site generation, and alternative energy sources. Such distributed resources will greatly reduce the risk of investments for both. On-site project investments may become part of a utility's investments.

METERING, BILLING AND DATA WAREHOUSING

SCADA and other metering technologies will be able to integrate with automated process and building control systems. Metering and verification for ESCO projects will be combined with billing. Simplicity of investment and ease of tracking ongoing results will be promoted as competitive advantages. The warehousing of energy use data will be used as a significant competitive advantage. With the data aggregated, on-site projects can be brought into emissions targets. Under the present, unaggregated arrangements most on-site projects are too small to cost-effectively participate.

BUILDING INFORMATION SYSTEMS

In the model of other globalized industries, industry leaders will increasingly become wired corporations, using information and communication technology to integrate internal processes, resources and linkages to external partners.

The information system serves as a network channel and also as a means of organizing knowledge. The utility could become an Internet content-provider. Users are drawn by content and good content providers are valuable.

The user-interface and operating environment must meet the needs without unnecessary complications. It could contain features such as personalized on-line research support and links to on-line energy markets. The more comprehensive the energy-related environment becomes, the more functions the customer needs to perform in it.

TECHNOLOGY DIFFUSION

Technology diffusion studies indicate that acceptance of innovation follows a pattern. Communication has been identified as the key to the process, starting with early innovators who are motivated and inclined to experiment. They tend to initiate the spread of positive messages. As this communication accelerates, the rate of acceptance becomes widespread for that market.

It is expected that this pattern will apply to changes in the energy business as utilities adopt an energy services model. This includes the adoption of information technologies as a way of linking utilities to service providers and to end-users. This should be viewed as the spread of an enabling technology and does not change energy-producing or energy-using systems. The change comes when utility and customer collaborations develop new processes and equipment during the final level of innovation.

Acceptance of the enabling technologies is the driving force for a potentially cascading series of new applications in distributed generation and end-user efficiency.

The information/knowledge system provides the framework for

access and interactivity. Traditional means of releasing information are being replaced with on-line conferences and Web networking. This allows a wide-ranging connectivity.

ENERGY MANAGEMENT SOFTWARE

Energy management programs address the areas impacting the indoor environment. These include energy, building maintenance and indoor air quality.

The software should be user friendly and allow users to work on an individual basis or use a wizard to guide them. The software should be modular so that new recommendations can be meshed into the existing software application.

The application software may be an OLE container like Microsoft Word or Excel. The OLE container is an application that consolidates the energy recommendations, which are separate applications called OLE servers.

Each server furnishes a bitmap in the client area of the OLE container application. Double clicking on one of the bitmaps displays a dialog box that prompts the user for data. The software prints a report based on the user's input. Changing the format of the report involves editing a Microsoft Word template file.

A software wizard will ask specific questions and depending on the answers, the wizard will embed only those servers needed. If the wizard is not used, the application will embed all servers and the user will choose from any of these.

Recommendations for lighting include lower lighting levels. See Figure 8-1 for the lighting levels recommendation. Double clicking on the lower lighting levels recommendation bitmap area launches the OLE server and displays a dialog box for data entry. Clicking the calculate button displays the savings and implementation costs. A report is printed by clicking the print button.

BUILDING MANAGEMENT SYSTEMS

A building automation system (BAS) can be used to automatically implement operation and maintenance strategies that lower costs and

Lowering unnecessary lighting levels - anticipated cost savings

Total number of lamps present   ___

Number of lamps removed   ___

Annual operating time (hours/year)   ___

Lamp wattage (Watts)   ___

Average kWH cost ($/kWH)   ___

Consumption cost savings ($/year)   ___

Demand charge ($/kW-month)   ___

Demand cost savings ($/year)   ___

Total amount savings ($/year)   ___

Estimated cost of implementation   ___

If the recommendation requires only a change in procedure there

is not implementation cost and the payback is immediate.

Figure 8-1. Lower lighting levels recommendation

enhance comfort levels. A BAS can provide other benefits such as increased operating efficiency and reliability, safer environments and better returns on investments. It can be an effective troubleshooting tool when used to collect historical and current data for analysis and documentation.

Rising energy costs are a growing concern for building owners. As buildings grow older, the operating efficiencies of building systems drop and energy costs go up. A BAS can improve operating efficiencies and eventually lower energy costs.

Energy management is a continuous process. Before an energy retrofit is implemented, a BAS can be used to develop a baseline of energy usage and environmental conditions. This makes it easier to calculate energy savings and compare environmental conditions to verify the performance of the retrofit.

Once a retrofit is complete, performance variables need to be monitored and adjusted to maximize the savings potential. The energy consumption is compared to the baseline consumption to calculate savings. The process can be done using a BAS.

Some building management systems can interface with other Windows-based applications, such as MS Word, Excel or Access. This allows the BAS to generate energy reports for analysis on standard programs.

A BAS can also be used to track operating efficiencies and alert the energy manager when efficiencies drop below a certain level. This includes automatic trending and reporting of a building's key performance indicators such as energy cost per square foot, energy cost per unit produced or energy cost as a percentage of net income. A BAS can link the operating efficiencies to the building's core business output.

A comparison of operating efficiencies between several floors of a building can be made. This type of checking can indicate that a piece of HVAC equipment, such as a chiller needs replacement or repair. Studies indicate that the operating efficiencies of a building can be improved by 10-20% using a BAS.

Depending on the equipment controlled by the BAS, there are several ways of using a BAS to lower energy costs. The BAS can conduct scheduled Start-Stop of light, HVAC, and process or manufacturing equipment. Automatic control of HVAC and process-controlled variables, such as temperature, static pressure, humidity and flow can be based on actual load conditions.

Automatic load shedding and load shaving can take advantage of cheaper power during certain hours of the day. This is a major advantage in a deregulated market. Equipment interlocking can be used to ensure that equipment consumes energy only when necessary. This can reduce energy consumption by 15-30%.

Integrating end-use metering into the building management system using evolving communication protocols is now possible. Systems are available that use networked power metering with multiple metering points on a single RS-485 network.

ELECTRICAL METERING EQUIPMENT

The approaches to electric metering include utility-provided information, third-party systems that tap into utility meter data and standalone metering systems. The Bonneville Power Administration Headquarters did an installation demonstration in its 7-story office building of over 500,000 square feet, with a peak demand in excess of 2-MW. Several systems were available with different levels of installation and operation (Table 8-1).

Table 8-1. Metering Systems

Provider	Description
Pacific Power (serving utility)	Mail monthly reports of interval data Provide pulse signal from revenue meter Provide access to revenue meter register data
TeCom	Hardware/software for collecting pulse and for collecting pulse and analog data
Energard	Hardware/software for collecting pulse data and temperature, incorporated into predefined reports distributed via Internet
Enerlink	Software only for access to utility register data
Schlumberger	Utility-type meter with pulse output along with software for accessing register data (used for submetering)

In order to get information from the electric utility meter, the existing standard electronic demand meter can be replaced with a meter with an internal modem. Some meters have DOS-based software to access the meter data. Others provide pulse outputs for direct access to the utility interval data.

The InterLane Power Manager from TeCom includes an Interactive Control Unit (ICU) which can accept nine pulse as well as eight 4-20-mA analog inputs. These inputs can be used to activate a warning device or can be interfaced with a control system. Data downloads and programming are done through a modem. This system has extensive analysis and graphing capabilities, including data analysis, load profiling, data normalization and user-created bill generation.

The Energard Envision system includes a Remote Processing Unit (RPU) which monitors pulse data from the building revenue meter, along with outside air temperature. Extensive reports are made available to the user on a Web site. These reports include real-time displays as well as energy tracking reports. Energy use is normalized based on temperature for the energy-use tracking reports. Setup is done through the phone line and the user accesses the reports using a standard Internet browser. Setup includes the characteristics of the site and defines the kWh and kvar that each connected channel will monitor. The user can then access the ICU via modem.

An important step in the setup is defining the pulse weight (kWh per pulse). Pulse signals, called KYZ pulses, are the standard output signal from a utility watt-hour meter. In the case of a revenue meter, this quantity is supplied by the utility. In the case of pulse signals from a non-revenue meter, the device is programmed, or preset, with a given pulse weight that is used to set the recording device. If the recording device is set with an incorrect pulse weight, the system will give incorrect values.

A changing load or changing interval also affects pulse-data. Pulse weights must be set to the proper value for a given load and given recording interval in order to achieve good resolution. For a 30-minute interval and a 400-kW load, a pulse weight set at 1-kWh per pulse will give a pulse count per interval of 200, giving fairly good resolution. If a user wants close to real-time data, and views the data at a 1-minute interval, there will be only eight pulses per interval, giving very poor resolution. If the load decreases to 100-kW and you use a 5-minute interval, the pulse rate will also be eight pulses per interval, giving poor resolution. Rapid fluctuations in a display can be interpreted as abnormal conditions, when in fact they may be the result of using a pulse interval that is too short.

Standard telephone wire for pulse signals is used for runs over 30.5

meter (100 feet). Software can be installed on a PC equipped with a modem and access to a phone line. The software has several protocols designed to communicate via modem with a variety of standard utility meters and recorders. This software package has the capability to generate a bill, but the manufacturer must create the rate files for installation into the software package. The package can also apply different rates to the same load profile. Users can graphically modify their actual load profile to observe the change in energy bills if they decrease peak demand or shift the peak to a lower rate period.

SUBMETERING LOADS

The systems were used to submeter the energy for a new 24-hour-a-day operation. Two primary feeders were added to supply the operation. Both were connected to an Uninterruptible Power Supply (UPS) and an emergency generator.

Utility-type electric meters were used with their pulse signals supplied to the TeCom system. Installation of a utility-type meter requires interruption of the power supply, so this installation was coordinated with a planned power outage. An electronic meter (Schlumberger Fulcrum) was used on one feed and an electronic meter with an internal modem (Schlumberger Vectron) was installed on the other feed.

The data showed that the 24-hour operation load was relatively constant. Programming the utility's rate schedule into the TeCom system revealed that the new operation cost about $2,000 per month.

Load profiles can be viewed from the 15-minute interval data. It was found that the minimum building load was about 50% of the peak load. One period can be compared with another. Overlays of one week to another and single-screen views of a month of daily load shapes can be used for load profile modeling.

Consumption data can be normalized with user-defined quantities. Weather-related adjustments can be made for a comparison of one time period to another.

Using a laptop, energy managers can dial into the building and check the real-time energy use. The software also allows access to Web site energy accounting.

There are future plans to meter two chillers in the building to provide continuous monitoring of electric use and chilled-water output. This will require kW meters, as well as temperature and flow measurements. The kW input will be monitored with kW transducers with an analog output that can be connected to the building automation system, as well as to the TeCom ICU. This information will be used to evaluate operating efficiency on a continuing basis.

UNBUNDLED UTILITIES

As deregulation unfolds, the vertically integrated utilities of the past can be expected to unbundle their business into the following components. Generating companies (GENCOs) will produce commodity bulk power. Their customers will include distribution utilities, end-use customers, and brokers. Transmission companies (TRANSCOs) will deliver wholesale power for the GENCOs and act as brokers.

Independent system operators (ISOs) will oversee transmission in designated control areas. They will try to ensure that buyers and sellers of electricity have access to the transmission, control the dispatch of generation and provide real-time balancing of load and generation. ISOs are more likely to function as quasi-governmental entities or not-for-profit corporations with government oversight. Power exchange operators (PXs) in some areas will provide common generation purchasing or pooling services. Services such as spinning reserves, frequency control and reactive power supplies will be provided to ISOs via competitively bid contracts.

Distribution companies (DISCOs) will deliver energy to retail customers in franchised service areas and will continue to be regulated by state utility commissions. Investments in infrastructure will be based on meeting demand for power while accommodating common standards for quality and reliability. Competition may eventually evolve to DISCOs undergoing periodic bidding to provide services in the franchise area.

Energy service companies (ESCOs) will provide value-added products and services ranging from power purchase services, on-site distributed generation and cogeneration, premium power quality and reliability as well as energy efficiency and demand side management.

POWER RELIABILITY

The North American Electric Transmission is constrained, and pushing it beyond its physical limitations causes reliability problems. The actual capacity available is a function of the lines and network. When the network limits are pushed, thermal overloads can occur along with voltages out of proper ranges and there may be instability from generators losing synchronization.

U.S. utility networks were designed to deliver generated power to nearby population and load centers and not across large transcontinental distances. The average network capability distance for U.S. utilities is about 200 miles. In the United Kingdom, competition in electricity supply was introduced years ago and power suppliers in the UK have an average network capability distance of 600 miles.

In the British Isles, the longest distance between generation and load is only one-fifth of that in the U.S. Also, there are more than 3000 independent utilities in the transmission system in North America, but in the UK, deregulation required the breakup of just one state-owned utility. Building a super grid in North America would require a massive construction program and that would have to be factored into the price of electricity.

OASIS

The federally mandated OASIS (Open Access Same Time Information System) is the electronic system which the ISO will use to service the spot market on the PX. Supply information will include the available transmission capacity and total transmission capacity. Transmission-service products and prices along with transmission-service requests and responses will be available as well as additional offerings and prices.

Regional variations exist among the different ISOs. In New York, existing power pools will use integrated PX and ISO functions. In California, separate PX and IOS entities have been created. Regional differences depend on the number of existing utilities that participate, historical operations and relationships and the geographic nature of the transmission infrastructure.

ANCILLARY SERVICES

In the past, powerplant operators concerned themselves with producing kilowatts, while control centers took care of support functions, such as spinning reserves, reactive power, black-start capacity and network stability. These functions are the generation-supplied components that add to transmission reliability. They are referred to as ancillary services or interconnected operations services (IOS). Under the federal open access rules, these services will be unbundled from distribution and generation services. Although the term ancillary services implies something subordinate or supplementary, ancillary services are essential to the reliable operation of the grid.

Ancillary services are being defined by an IOS Working Group formed in the U.S. and supported by the North American Electric Reliability Council and the Electric Power Research Institute. There are 13 separate services that may be supplied, depending on generation capability and generation characteristics. The three main areas are reactive power, black start and power reserve.

Accounting for the cost of ancillary services is critical. In the past, this was done by each utility as it covered its native load, or was organized by an Area Control Center to cover the load inside a defined area. The variable cost of the service was not calculated, it was rolled up or bundled into the overall cost of producing electricity.

NEW TECHNOLOGIES/DISTRIBUTED GENERATION

In the area of ancillary services, the emergence of Distributed Generation (DG) is prominent. These are generation units in the 0.5 to 10.0-MW range. Networking or pooling of these resources produces bid blocks that can compete in the regional PX. The capabilities and response of the DG pool will not be distinguishable at the ISO level from combustion turbine or central generated power.

Distributed generation can address all areas of ancillary services better than central plants. Distributed generation is easier to site, lower at first cost and when properly designed can have a positive impact on air quality issues. It can provide a faster, more measured response compared to a central plant. Since distributed generation is deployed on the demand side of the grid, it relieves transmission and distribution requirements. Besides the faster response time, there is the ability to inject kvars on the demand side of the system.

Other technology innovations will transform how energy is generated, delivered, and used. These are often called Distributed Resources. Improvements in technology have decreased the importance of economies of scale. It is no longer necessary to build a 1,000-megawatt generating plant to achieve generation efficiency. New gas turbines as small as 10 megawatts can be efficient. Diesel and gas-fired internal combustion engines make distributed generation and cogeneration increasingly feasible today, while advanced fuel cells and micro-turbines offer additional potential in the future.

Along with energy storage technologies, these distributed resources provide alternatives to the traditional central generation facilities. The U.S. Energy Information Agency estimates that these smaller generation technologies will make up 80% of the more than 300 gigawatts of new generating capability to be added by 2015.

New technologies will also determine how electricity is delivered and used. Computers allow monitoring of the distribution system with real-time information. They indicate where the load is, where the crews are and what switches are open on a desktop computer. More sophisticated switches in the field can automatically report problems and allow the restoration of service more quickly after an outage. This provides instant readouts of equipment status and automated reports of voltage fluctuations and dropouts.

Innovative sustainable generation is becoming increasingly important. New advances in wind micro-turbine technology and photovoltaic solar power makes these older green technologies more feasible. Other technologies include advanced fuel cells and biomass generators. Energy storage technologies, such as advanced batteries, ultracapacitors, and flywheels, offer new opportunities for demand-side management as well as hybrid renewable power systems.

As states around the country allow retail customers to choose their electricity provider, industries of all types and sizes are becoming concerned with the question of preparedness. A company needs to be prepared and understand the upside and downside risks.

OUTSOURCING

Outsourcing is the process of turning over a function or activity to another entity specializing in that function. Outsourcing involves the contracting for services other than core company activities. Reducing the primary company's administrative costs and performing the function at a reduced rate are the basic advantages of outsourcing.

Outsourcing is not a new concept, but in the last several years it has gained momentum in certain industries. Many companies today use some form of outsourcing in their current operation. It may be mail services, human resources, relocation, travel, information systems, security, maintenance or accounting. The Information Technology and Information Systems sectors have been major users of this concept.

The automobile industry uses outsourcing for many of its parts and assemblies. In a cost comparison for car seats, the average labor cost by Ford or GM is about $45. Parts suppliers can do this for about $27.

Credit card operators have used outsourcing to manage billing and data management functions. Other companies have used third parties for the manufacturing of products. The primary company continues to manage the sales and marketing while the third party is responsible for fabrication or manufacturing.

Pharmaceutical companies are even using third parties for research and development of new drugs. The ability to use third-party research laboratories allows these to concentrate on marketing and governmental approvals.

There will be greater efforts to initiate an outsourcing activity of a dollar intensive function such as energy. Companies that do not have complete energy expertise in-house will seek outside help. The energy industry is becoming too complex for part-time energy management. Even at small plants, the Plant Manager, who has been performing the energy procurement tasks in the past, will no longer be able to keep up with the changes in the energy industry well enough to be able to make informed decisions. The risk factors will escalate as the energy industry becomes more competitive.

Energy outsourcing involves obtaining information on energy commodities and establishing a source for reference. This may be an in-house effort or involve an energy advisor. Historical records of energy usage for all areas are needed. These are evaluated for prioritizing and for identifying high cost and potential problem areas.

An in-house energy team will normally include purchasing, production, finance, engineering, middle management, consultants and outside contractors. The team should establish goals, develop strategies, and obtain management approval.

The team needs to be informed about the laws and regulations applicable to their sector. This may be obtained through attorneys, energy advisors and group associations involved in energy activities.

A Request for Proposal (RFP) can be used to solicit proposals from energy outsourcing firms. The RFP can be used for learning about any new options available and to confirm that conventional options are available at competitive prices.

The development of the RFP may require the services of an energy advisor in order to focus on the appropriate objectives, to identify qualified outsourcing companies capable of performing the function, and to properly evaluate the proposals, objectives, and contract business terms.

Each aspect of the ongoing energy activity must be reviewed, and goals and objectives revised to comply with changing laws, regulations and company policy. The economic aspects must be evaluated on a continuous basis to ensure that the program is working.

What to outsource depends upon the nature of the primary company, its core business, and its expertise. Energy commodities include electricity, natural gas, coal and fuel oils. Activities include procurement, contract negotiations, bill review and reconciliation, general management and reporting functions.

ENERGY ISLAND/ENERGY CENTER

The new technology includes the Energy Island and the Energy Center concepts. These two concepts involve the conveying of physical energy-related assets to an outsourcing company. In both concepts, the outsourcing company conducts operating efficiency studies and improves the efficiency of the equipment, the operation and the overall energy supply chain inside the facility. The Energy Center is similar to the Energy Island concept except that it also includes new generation equipment in addition to the efficiency improvements of the Energy Island. In both concepts the company is free to direct its capital toward core business activities.

Another advantage of the outsourced function is the ability to aggregate energy usage at multiple sites. This can be a significant advantage as the electrical power industry is restructured.

COMPUTER SYSTEM INTEGRATION

The state-of-the-art is such that a high level of integration is becoming a commodity, in standard software packages. The longevity of equipment means that controllers and software do not always reflect the latest technologies, especially in the area of communication capability. Often, replacing expensive equipment just to improve the controller capability is not practical. While PCs are recycled quite rapidly, equipment controllers and the underlying designs of the controller, may be 5 or more years old which is aged in computing years. Enhanced communications capability can justify adding features to existing equipment, such as tracking energy or fuel use.

POWER MONITORING SOFTWARE

There are software packages that allow businesses to monitor and track energy use down to individual computers, lights and other devices. Providers of such software packages include Energy Company and Power Measurement. This software can provide savings of 5 to 20% on gas and electric bills.

In the past most energy-monitoring systems were used by giant retail chains and industrial concerns. Now, these solutions are becoming available to smaller companies with an annual energy bill of $100,000. Energy deregulation is the driving force.

Silicon Energy is a provider that was founded in 1997, it monitors a company's meters down to the circuits that control heating, air conditioning and other energy in the buildings. It calculates and posts the data on-line, where it is updated about every 15 minutes. Using a Web browser, clients can view the energy usage at peak times and match that data against often complex rate plans. In some cases, there has been the discovery of buildings that have the air conditioning running in the middle of the night, and buildings where heating and cooling are going on simultaneously.

Energy managers can be automatically paged as usage levels approach critical thresholds. The system can be set up to respond automatically or allow company employees to power down themselves. Actions include turning down air conditioning by a few degrees for an hour or two, readjusting heating and ventilation setpoints or turning off noncritical lighting in the late afternoon.

In an office environment, computers and computer networks are too important to shut down. But, you can reduce the temperature by a few degrees on the cooling system for an hour or shut down large water heaters for a short time.

Businesses like laundromats, cold-storage facilities and even small in-and-out grocery stores often run heating or cooling and refrigeration around the clock and consume a lot of energy.

In the Power Measurement system, the data are collected from small sensors attached to the power meters and sent to the computer, which graphs the data into snapshots that can be used to make energy decisions. Depending on the number of meters, the system can measure loads from parking-lot lighting to ventilation systems.

Many utilities companies in deregulated states where users can choose their energy providers, will be providing such monitoring solutions directly to customers as a retention tool. Energy Interactive is selling its on-line, browser-based solution to regional utilities, which then pass on to commercial customers for free or for a small rate.

One company that serves the Washington, DC, area offers the service to businesses for $16 a month. But they must have meters on-site or purchase additional monitors for more detailed monitoring which start at about $500 each.

Energy Interactive sells a software package companies can use with or without the company's on-line service. Users collect the data, which utilities often provide on request, and load it onto the desktop application. By viewing the usage charts on-screen, a company can determine how much they will save or spend per month, depending on energy decisions they make. The software package costs about $2,000. This information can be used to negotiate favorable rates and adjust energy usage to optimal levels.

ADDING METERS

A system that lets you measure usage on a per-machine basis can be powerful, but it may mean buying more meters for the software to monitor. Power Measurement sells its own meters, but Silicon Energy leaves that to companies that have been doing it for years, like Honeywell. Prices for meters range from a few hundred dollars to a few thousand, depending on the features required. Smaller companies can also purchase meters to measure the energy consumption of their biggest or most wasteful loads and adjust their use without software.

BUILDING MANAGEMENT SYSTEMS

Building management systems can meet most control needs including problem areas in an existing building to a total system for a major complex or campus. Capabilities and features of building management systems include:

- scheduling for weekdays and holidays,
- after-hours tracking,
- management reports and logs,
- interactive graphics,
- direct digital control,
- comfort monitoring by zone,
- energy-saving software, and
- simple operator interfaces.

Total integrated systems include HVAC, controls and building management. A building management panel on the monitor screen is used to coordinate the control of HVAC equipment and related building systems. It provides the user with management information and networking with packaged HVAC equipment, zone controllers and others on a communications link, creating an integrated system.

A system like the Trane Tracer provides building automation and energy management functions through stand-alone control of HVAC equipment. It offers direct digital control (DDC) capabilities, and full monitoring and control of conventional HVAC equipment. The system provides a graphical user interface, high-speed communications and distributed intelligence.

Advances in computer technology include local area networking, powerful operating systems and object-oriented databases. These advances allow the system to provide valuable diagnostic data from unit-mounted controls on the HVAC equipment.

NETWORKING

Plant engineers and facilities managers need an understanding of networking and systems interoperability. Facilities engineers and managers are trained in disciplines related to facilities engineering, operation, maintenance and management. Communications and systems interoperability issues are outside of their main focus and the entire communications industry is evolving at an extraordinary rate.

Most automation systems and networks have been developed over time. The issues of interoperability between installed systems and future upgrades can provide a challenge. The goal is to provide solutions that will lead to further development of facilities automation without obsolescence of the already installed systems.

Systems interoperability is a complex task and end users who underestimate this complexity can make costly mistakes. Several basic definitions are related to systems interoperability and networking.

COMMUNICATION PROTOCOLS

To communicate from one computer to another, there must be connectivity and interoperability over a network. To transmit, receive, interpret and acknowledge messages over the network, it has to be in a format the computers or controllers can understand. The rules and procedures for data transmission over a network are defined in the form of a computer protocol.

Protocols may be proprietary, open or standard. Proprietary protocols are developed by systems or computer manufacturers to communicate on hardware and software supplied by them. They are designed and tested for vendor specific systems. Proprietary protocols are not openly published and other systems or computers cannot coexist or communicate with the vendor specific system or computers on the same network.

The proprietary nature of these systems provides systems reliability and integrity, but it precludes the end user from utilizing other vendor's off-the-shelf systems on the same network.

Proprietary protocols have been widely used in the building automation industry. They still have their share of the market even with the growing appetite for systems interoperability. The responsibility for the automation systems performance as well as for systems communication is with one vendor.

Many open protocols start out as proprietary protocols. They become public domain and allow other system developers to write interfaces and share data on the same network. Opening up proprietary protocols means disclosing the procedures, structures and codes used by these protocols for other systems to communicate on the same network.

The best of the open protocols become defacto, or industry standard protocols and are widely used by end users and system integrators. They have been tested in many applications and refined with many improvements over time.

Most DDC systems on the market today have open protocols. Many vendors offer interoperability with systems related to the building HVAC industry. Problems associated with systems integration based on unproven and untested drivers can still exist.

Standard protocols for building controls include the BACnet protocol for building automation and control networks and the European Profibus developed for building automation systems in the European market.

OSI MODEL

Most of today's protocols are based on the 1977 model for Open Systems Interconnection (OSI) published by the International Standard Organization (ISO). OSI is a model, not a protocol, adopted by the computer industry which groups the activities needed for data transfer over the network.

The OSI model defines the activities related to communication protocols in seven layers. The highest layer is concerned with the application program, while the lowest is the physical layer. It is concerned with the network media, such as twisted shielded wire or fiber-optic cables. Most communication protocols are based on the OSI model, using all or some of the seven layers (See Table 8-2).

Table 8-2. OSI Model

Layer	Function
7	Application
6	Presentation
5	Session
4	Transport
3	Network
2	Data Link
1	Physical

In the OSI model, each layer has an associated address, and processes data delivered from the layer below or above, or from the same layer of another node on the network. The upper software layers deliver exported data to and from the application software. The lower layers provide the data export and import over the network.

The Physical Layer specifies the communication path or physical media of the network. It defines the transmission media, such as twisted pairs, coax cable or fiber-optic cable along with the electrical signal levels and the characteristics of the signal, such as transmission wire lengths and the speed and amount of data it can handle without repeaters for signal regeneration.

The Data Link Layer controls access to the Physical Layer and network media. Its main function is control, sequencing and synchronization of the data along with level error detection/recovery and other functions for control of the physical media.

Several IEEE standards comprise the Data Link layer:

- High Level Data Link Control (HDLC),
- Logical Link Control (LLC),
- Medium Access Control (MAC).

HDLC and LLC are typically implemented in software while MAC and the Physical Layer are implemented in hardware, typically as PC interface boards. The hardware provides bit handling, encoding, error detection and recovery, address detection and recognition.

Other IEEE standards used for networking include: 802.1 High Layer Interface, 802.3 CSMA/CD (Carrier Sense Multiple Access/Collision Detection), 802.4 Token Bus, 802.5 Token-Ring and 802.6 Metro-area Network Media and Media Access.

The Network Layer provides the interfaces between the Transport Layer above, and the Data Link Layer below in the OSI model. The Network Layer is required to establish and terminate connections between the originator and recipient of information. The Network Layer assigns addresses (numerical codes) to each computer node on the network. The addresses also identify the beginning and end of the data transmission. The Network Layer delivers the information directly to its destination or finds alternate routes.

The Transport Layer is concerned with reliability and data integrity. This layer provides error recovery, by requesting re-transmission of incomplete or unreliable data. Standards associated with this layer include the X.25 wide area protocol and the Transmission Control Protocol (TCP).

The Session Layer sets up and terminates the sessions between application programs on the network. It synchronizes the data exchange for network communications.

The Presentation Layer's main task is to convert data received from different protocols to a format that can be utilized by the application program. It provides data formatting, translation and syntax conversion. This layer is responsible for receiving, unpacking, decoding and translating the data for the application program.

The Application Layer provides services for the application program. These services include data transfer and management, file access and remote device identification.

COLLAPSED MODELS

Most automation systems protocols, such as BACnet, use a collapsed OSI model, where individual layers are collapsed into equivalent layers, or left out. BACnet uses an Application Layer, Network Layer, Data Link Layer and Physical Layer. The two lower layers have 5 interface options: Ethernet, ARCNET, RS-232, RS-485, and LonTalk.

The operator interface provides global control. It serves as a communication link between the operator and the building management screen panels.

A graphics-oriented building automation system can provide the monitoring and control needs of large buildings. General-purpose controllers provide the capability to tie the equipment into the communication network. They include thermostat control modules and programmable control modules.

Programmable Control Modules (PCM) provide direct digital control and monitoring for a wide range of HVAC and other applications. These uses include controlling air handling equipment, interfacing with water chilling units and boiler systems and controlling pumps and cooling towers. This provides optimized monitoring, control and diagnostics of chillers.

Chiller control products include chiller plant managers and DDC chiller sequencers. The comprehensive monitoring and control of chillers includes the responsible and safe handling of refrigerants.

A chiller plant manager building management system provides building automation and energy management functions through stand-alone control. This includes the monitoring and controlling of HVAC equipment, providing the user with management information and networking with other systems.

A direct digital control chiller sequencer is a sequencing panel that provides start/stop control of several chillers and associated pumps, setpoint control, lead/lag, soft loading, failure recovery, chilled water reset, ice building and alarming.

MAINTENANCE MANAGEMENT

One study by the American Society of Heating, Refrigeration and Air Conditioning Engineers (ASHRAE) indicates that over the life of a building, 50% of the building costs can be attributed to its operating costs. Maintenance costs make up a significant part of the building's operating costs. Implementing maintenance strategies can help to improve operating efficiencies and maintenance costs may be significantly reduced. The proper maintenance strategy can improve the building's operating reliability and impact staff productivity. High operating efficiencies, low operating costs and high productivity are all related.

Maintenance technology has evolved over the last few decades. In the past, maintenance management was primarily reactive, this meant that equipment was not repaired until it failed. Later preventive maintenance evolved, which defines time-based tasks for equipment.

PREDICTIVE MAINTENANCE

Recently, the use of predictive maintenance has been in an era of growth. This type of maintenance allows building managers to identify and detect problems before they cause failures. An example would be oil analysis on machines. Technology now makes it possible to detect the source of problems. This is also called proactive maintenance. Examples of proactive maintenance include vibration analysis on fans and pumps, which help to detect problems such as improper balancing or misalignment. These newer service strategies may not always be appropriate for each unit of equipment. An optimum maintenance management process involves choosing the appropriate service strategy for different pieces of equipment, based on equipment redundancy and potential damage.

This process is often referred to as Results Oriented Service because the service strategy for every piece of equipment is selected based on business objectives. This strategy is also called Reliability Centered Maintenance.

In reactive maintenance or proactive maintenance, a BAS can collect and analyze critical data and then take the appropriate actions. A BAS can indicate when a problem is detected. Notification can be an on-screen visual alarm, a text message sent to a pager, a work order in a computerized maintenance management system, a phone call or an e-mail message.

Alarm management is a feature of the BAS. Notification can be based on how long the equipment has been running. This is time-based monitoring for preventive maintenance. The strategy could be based on the condition of the equipment.

In this type of condition-based monitoring for predictive/proactive maintenance, if the BAS senses an unacceptable vibration level in a chiller that supports a data center, the BAS can notify someone of the problem, before the problem gets worse and causes an unscheduled downtime.

This type of maintenance strategy increases equipment life, reduces operating costs and increases staff productivity. Building owners have realized maintenance cost savings of 10-40%. Other benefits in using a BAS for maintenance management include an increase in system reliability and system efficiency which results in longer equipment life.

PROACTIVE MAINTENANCE

The use of computers and low-cost sensors allows degradation diagnostics. A computerized real-time picture of the problem is possible. The solution can also be computer generated and sent simultaneously to the operations, maintenance, engineering, and administrations staff. Asset management can make informed decisions based on known conditions, defined degradation rates and, in many cases, accurate estimates of equipment life (prognostics).

Over the life of the equipment, the savings provided by this proactive approach are estimated to be 5 to 10% above even the predictive approach, including the initial investment costs. In a balanced approach using reactive, preventive, predictive, and condition-based maintenance it is possible to generate a total production life-cycle cost savings of 25 to 40%.

In 1990, the United States Marine Corps and the Pacific Northwest National Laboratory (PNNL) demonstrated the effectiveness of this emerging technology. The Marine Corps produces thermal energy (heating and cooling) for its base facilities using central energy plants.

The central heating plant at the Twenty-nine Palms, CA, base is a gas-fired 120-MBtu/hour pressurized hot water plant that provides thermal energy (heat) for 20,000 Marines.

The effort resulted in a set of on-line computerized, operations-oriented tools. This provided the operations crews with point and click information access to the processes at the plant, system, and component levels. The safety and the efficiency of the components and the process are monitored, and root cause solutions are generated.

To obtain an accurate record of change, a detailed baseline characterization was performed. This included not only the common metrics, such as plant overall operating efficiency and maintenance machinery repair records, but also other functions that must be integrated to provide the infrastructure required for continued operations.

The project increased plant thermal efficiency by 17% and reduced the gas bill by over a quarter of a million dollars each year. A concurrent increase in plant available capacity eliminated the need for a fourth generator unit, saving $1,000,000.

A life-cycle cost study was also performed. It was found that the base was saving approximately $480,000 per year on its life-cycle heat plant bill and would pay back the original research investment in 4 years.

The USMC incorporated this technology into its Energy Conservation Campaign Plan and is applying an advanced version of the technology to the thermal and electrical generation systems at its Parris Island, South Carolina, cogeneration plant.

This software was designed to provide the operations staff with on-line information to allow precise control of the combustion, heat transport and hydraulic processes at a central heating plant. Several problems can exist at this type of plant. Equipment problems identified by the operators can be repaired, but the source of the problem would be unresolved, so chronic plant problems continue to be unchecked. Machinery material deficiencies noted by the maintenance technicians may not be investigated and the machinery can continue to break down from the same cause. Plant administration may not know the plant performance level or what resources are necessary to maximize plant life-cycle efficiency.

The different areas that offer solutions to these problems include:

Performance monitoring—provides the basis for understanding, reporting and improving plant efficiency and reliability,

Procedures—ensures procedures provide appropriate direction for plant efficiency,

Plant modifications—provide solutions to problems or upgrades for higher efficiency for the plant process,

Support organization—ensures effective implementation and control of plant engineering support needs,

Document control—ensures documents provide accurate, as-built information sufficient to support plant requirements.

If any of these areas is even partially dysfunctional, the impairment is transferred to some degree to the efficiency, reliability, and safety of the entire plant.

Performance monitoring involves: instrument calibration frequency determination, component, system and process testing, process performance baseline analysis, performance analysis, machinery reliability analysis and process performance reporting.

These elements are needed to accurately determine the adequacy of the performance of the process. A task such as machinery reliability analysis requires detailed listings of maintenance data on failures and resource availability information. The level of detail of these tasks may be less explicit in small plants, but the task elements must still be performed.

The USMC project at Parris Island extended the integration beyond thermal and electric generation at the site and included demand side management and waste treatment facilities. Data generated in the central energy plant, building energy management and control system (EMCS), and alarm information from the site waste treatment plant were routed to a central database.

The data are displayed in a form customized to the users' need. It includes operator information, status and failure information, design and performance data, and administrative plant efficiency and asset management data. This type of structure provides each of the plant functions with the data needed and allows other functional disciplines to view the plant through the perspective of their coworkers. This provides important cross-training benefits.

The actual training function is physically decoupled by requiring that it be performed in a different, non-networked computer. The potential for collisions between a training session and real plant operations cannot be left to chance.

COMFORT LEVELS

Occupant comfort is important and a BAS is an effective tool in maintaining comfort levels in a building. It can monitor and control the space temperature, pressure and humidity from a centrally-located computer. The BAS offers the flexibility to be responsive and automatically adapt to the occupancy levels at various hours of the day. Customized reporting by a BAS can help to create a comfortable environment for the building occupants.

A BAS can be set up to generate customized, management-by-exception reports that display the information needed to assess the comfort levels in the building. These management-by-exception reports can be used to identify problem areas.

The BAS could check if the air-conditioning unit fan is on. If the fan is on, then the BAS could check the compressor. Based on the diagnostics, it can identify the source of the problem.

An environment that does not address occupant and equipment safety is not optimum. A BAS can be integrated with other building systems, to enhance the safety levels in the building. Multiple building systems can be integrated with the Building Automation System for improved safety and security.

IMPROVING SAFETY

Many of the automatic responses to a fire condition are set in the local codes. This includes turning off the air handling unit on the floor of incidence and homing the elevator. The types of responses required by the fire codes are under the direct control of the fire alarm system. Other building systems can provide a secondary response.

The secondary responses, that can be coordinated by a BAS include turning on lights on the floor of incidence to aid evacuation. Air handling units on the floors above and below the floor of incidence can be activated to contain the migration of smoke. This is sometimes called a sandwich system.

Message control includes inhibiting HVAC alarm reporting on air handling units turned off to limit distraction and sending audible evacuation messages based on the occupancy of the zone. The system could also send textual and graphical evacuation messages to users of the office automation network at information kiosks and other public displays. Unlocking security doors can be done to aide evacuation.

Inhibiting alarm reporting on security doors (fire exit doors for leaving the building) should also take place along with printouts of lists of occupied zones and floorplanes showing fire devices and exits.

PLANT SECURITY

Many security systems use a security alarm trigger for local CCTV cameras with displays on a dedicated monitor and recorded on a VCR. Locking out the elevator operation for the floor in response to a security alarm can delay the intruder. Another response to a security alarm is to turn on all lights in the area. Turning on lights in response to a security alarm can limit damage by intruders as well as protect the guard sent to investigate. Local CCTV cameras will also be able to record a better image with the lights on.

BUILDING HEALTH

A BAS can be used to monitor a building's health and make decisions to prevent the building's health from degrading. Indoor air quality (IAQ) is important and has received more attention in recent years.

The American Society of Heating, Refrigeration and Air Conditioning Engineers (ASHRAE) has a standard for indoor air quality, (ASHRAE Standard 62-1989, Ventilation for Acceptable Indoor Air Quality). One of the key parts of the standard is the amount of fresh air required per person for various types of building environments. For office environments, the minimum amount of fresh air is specified to be 20 cubic feet per minute (CFM) per person. This can be used to determine if the quantity of fresh air being introduced is adequate.

One way to check the percentage of fresh air in the supply air stream is to measure the Carbon Dioxide (CO_2). Using CO_2 sensors, the level of CO_2 can be measured in a zone along with the supply air stream and the return air stream. If the CO_2 level in the return air stream exceeds 800 ppm (parts per million), an output signal can be sent to the fresh air damper to allow more fresh air into the building. The required fresh air CFM can be tied to the occupancy level. This will lower the amount of fresh air being brought into the building during unoccupied hours and can save cooling/heating energy. This is called Demand Controlled Ventilation and can be controlled with a BAS. It can improve the building's health by providing the occupants with adequate amounts of fresh air and helps to achieve energy savings.

CHANGING BUSINESS NEEDS

Ways of doing business are changing and buildings need to support these changes. One of the more significant trends is a shift in occupancy patterns. In the past, building occupancy tended to be a binary function; fully occupied or fully unoccupied. In the future, buildings will be spending a much greater percentage of the day in partial occupancy. Flexible hours are the main reason for this change. Energy usage is another driving force along with globalization and increasing competitive pressures.

Today's buildings need to adapt to the current occupancy level. Safety, security, HVAC and lighting can be zoned and controlled to accommodate partial usage of the building. There must be enough granularity in the zoning of these building systems.

Building systems need to be integrated to provide a total building response to the current building usage. Integration may involve time scheduling with a default time schedule for each zone that can be overridden.

Card access provides an input to the system so entry to the parking lot or the building advises the other building systems that the zone associated with the card holder is now occupied. Motion detectors using passive infrared technology can be used as occupancy sensors or security devices. Door lock indicators, which can be microswitches activated by the latch set can also be used as input to the security system or the building system.

Manufacturers also provide telephone interfaces to their BAS. When occupants want to change the occupancy modes of their zones, they can call the BAS and a voice prompting system allows them to enter passwords, current location and specific requests. The telephone interface also allows occupants to change local temperature setpoints.

A facility booking system can be used to reserve conference rooms and other shared resources. The booking information is automatically passed to the building management system.

When a zone is occupied, building systems can be programmed to operate the system controls. The sensitivity of smoke detectors can be set to "low" to avoid unwanted alarms due to dust and smoke generated in the space. Security devices in the zone can have alarm reporting inhibited.

Temperature setpoints are switched to the occupied level and ventilation requirement (CFM per person of fresh air) are set to the occupied level. Lights in the zone can be turned on.

Web browsers can also be used for building systems control. Web-enabled control can be done through limited access via the company Intranet. Some building systems also accept electronic mail messages requesting overtime usage of space.

WEB TECHNOLOGIES

Improved energy management means a tighter integration of the entire facility. Information integration at this level requires the ability to link systems of different types, providing information in various formats throughout the facility and often with geographically dispersed plants and offices. Until recently, this was a difficult and costly problem, but the increasing use of technologies that have matured on the Internet now provides cost-effective solutions.

The Internet has successfully unified a variety of computers, including mainframes, midranges, workstations, and PCs. In many businesses, the move toward electronic commerce has meant a standardization around Internet technologies, including Ethernet, TCP/IP, and the Web. These technologies are used to track not only their facility but also the status of suppliers including energy suppliers.

BUILDING AUTOMATION AND THE INTERNET

Building automation systems that work with the Internet take building automation systems on-line. An interoperable BAS system can send out signals through the Internet to a PC that the HVAC equipment is not working properly. The problem could then be acknowledged and corrected remotely. Weather information could also be gathered from the Internet and downloaded into a building system, which could then adjust the buildings' systems accordingly. The Internet connection could also be used to check utility prices across the nation on a regular basis and make adjustments according to a building's needs.

Using Internet technologies on a local network to create a local Intranet allows the integration of many systems without a significant impact on operation. Networking allows the display of new information, as well as the addition of new control features.

The Intranet servers can be Web servers with off-the-shelf Web browsers, such as Netscape Navigator or Microsoft Internet Explorer, combined with Java-based technology to provide better access.

The use of Web techniques is fairly straightforward. Information is provided to clients by a Web server. On the Internet, a server may receive many millions of requests for information every day and Web servers are typically dedicated high-powered computers. For a facility Intranet, the number of requests for information is relatively small, so a Web server can run on a standard off-the-shelf PC system equipped with a network interface. These systems can run the user interfaces. They can also function as real-time controllers.

Web enabling equipment allows anyone with a Web browser and the appropriate permissions can access the information. This makes access to control information uncommonly simple. Once the client computer is set up and networked, you just need the familiar Internet address, or URL. This address might take the same basic form as an external URL, such as www.ibm.com. Internally, an address such as heatingsystem/status.html might provide access to the status of a local system.

In a typical Internet Web site, most of the information is static since it does not change until someone physically changes it. Dynamic applications on the Web need to advance this concept by displaying rapidly changing information in near real time. To create, transport, and view this information, the Java language could be used. The Java applets are transported across the network and are executed on the user's computer inside the Web browser. The user interface looks familiar to anyone accustomed to Web Surfing. Since Java and Intranets mainly support only open protocols, this interface is usually widely available throughout the network.

OBJECT TECHNOLOGY

One flexible technique used to communicate across a network is the distributed object concept. Distributed objects allow software running on multiple machines to interact almost as if it were on a single machine.

Distributed object technology allows the processing and data to be distributed to various devices where it may be most needed or most effectively handled, including distribution to the Web applications.

Distributed object standards include the common object request broker architecture (CORBA), guided by the Object Management Group and the distributed component object model (DCOM) which supersedes another earlier Microsoft standard (COM) component object model.

CORBA and DCOM provide similar capabilities for large-scale and facility integration. For the users of Internet technologies, the object request broker that enables CORBA, has the advantage of being available on almost every computer platform and operating system. This includes the embedded real-time operating systems used to control machines. DCOM is available only on Microsoft Windows products.

Java provides the networking support for socket programming. A distributed object technique called remote method invocation allows clients using browsers to easily interface to servers supporting these standards.

Most embedded and real-time computer platforms support Java in the form of Java Virtual Machine (JVM). This is an application or library that converts the portable Java byte code into platform-specific code as it is executed.

Multiple protocols can run on one controller. These Web-enabled components can run on local Intranets using a choice of open communication protocols.

WEB ENABLING

If a device has RS-232 communication capability, a terminal server can be used to convert the RS-232 communication to Ethernet. Terminal servers are available from companies such as Lantronix, Kanematsu and Sierra Monitor. These systems can also provide some protocol-conversion capabilities, such as from MODBUS RTU to MODBUS TCP. Some terminal servers are also Web enabled. They allow user-supplied Web pages and applets to be downloaded to the terminal server and accessed from there. These applets can communicate with the terminal server and then to the equipment over the Ethernet connection, for monitoring, control and other functions.

One use for these applets is to provide tasks such as configuration, troubleshooting, and maintenance. The manufacturer provides these resident applets. To configure the system, you point a Web browser at the Web server in the equipment or terminal server and configure the system from the pages shown. Troubleshooting can be provided as lists of alarms or fault indications with suggested corrective actions.

The maintenance and troubleshooting can be handled remotely over the Internet. The use of the remote user-interface can provide information on all aspects of the machine or equipment. The hardware interface provides monitoring of the system and could also control the equipment. Capabilities well beyond the scope of the original controller can be incorporated simply.

There are many PLCs that now include some capability for embedding Web pages and applets within the controller itself or within another module on the controller such as an Ethernet communications module. Future advances in PLCs will enable support for real-time, server-side Java. This would allow the PLC to provide monitoring and real-time control. Web-based interfaces and integration could be provided directly from the PLC, which would program such logic in Java on both the client and the server.

Using the technologies of the Internet on a local Intranet can provide a cost-effective method of improving integration. Many new controllers may include embedded Web pages for troubleshooting and maintenance.

The new interoperability provides the ability to combine products from multiple vendors into flexible, functional systems without the need to develop custom hardware, software, or tools.

WIRELESS TECHNOLOGY

Wireless technology will be prevalent in the future. One study predicts that in a few years, more Americans will access the Internet from their cellular phones than from PCs. Another study predicts that 111 million Americans will use mobile data services while another firm predicts that 1.7 billion people worldwide will subscribe to wireless services with 245 million of them in the United States.

It is likely to be a major part of building automation. There are now smart vending machines that are equipped with wireless modems that notify the warehouse when stocks are depleted, eliminating the need for costly, on-site time-intensive inventory checks. Portable ticket stations, which are terminals linked to printers via wireless modems, are used in many airports.

SBC Communications, which used to be known as Southwestern Bell has embarked on one of the largest deployments of wireless-enabled notebooks. SBC will have over 30,000 service technicians equipped with customized notebook computers from Panasonic. The company expects the change will reduce each service call by 10 minutes and allow an additional service call each day. Custom components include a line tester that fits into the notebook's device bay and an internal 56K analog modem. The built-in line tester eliminates the need for a stand-alone tester. The system allows access to all the plant data and can find where all the facilities are. There is an integrated testing capability for troubleshooting and finding line problems.

The type of wireless technology used depends on the maximum distance between the home base and the remote site. When you are roaming a building, wireless extensions, such as bridges or routers, can link you to a local area network. To cover a wider area, such as a campus or building complex, a series of LANs using wireless extensions can be used to transfer any kind of data. Infrared or low-power radio equipment linking PCs to portable computers can be used. Infrared devices can also be used to create wireless LANs, which can connect PCs on a peer-to-peer network or connect a node to a LAN.

Wireless technologies based on radio frequencies includes cellular technologies such as cellular digital packet data (CDPD) which is a digital alternative to analog-based cellular communication technology. Wireless can handle any information that can be transmitted over a coaxial cable or a telephone line.

One of the oldest wireless technologies uses radio frequency (RF) signals, which send and receive data using low-power transmitters and receivers. The range for local-area RF equipment starts from 30 feet to about 100 yards and some equipment can reach 600 feet to 1/4-mile (1,200 feet).

Most local-area RF equipment operates in the Industrial, Scientific, and Medical (ISM) bands, which covers radio frequencies that do not require a broadcasting license. The equipment must meet certain power and bandwidth restrictions enforced by the FCC (Federal Communication Commission).

Low-speed devices, range from about 115-kilobits per second (Kbps) to 250-Kbps while the high-speed devices can transfer data at 1.25-megabytes per second (10-megabits) or greater, which is about the same speed as Ethernet.

Some wireless connections include RF transceivers that attach to the serial port on the PCs. These devices transfer data at the rate of the serial port within about a 10-meter range. These small card-sized transceivers can transfer files between a desktop PC and notebook PC. They form a two-node network in which one machine, typically the portable device, can access resources available to the other unit, such as a printer.

Infrared transceivers can also plug into a serial port and transfer data between ports. This allows the infrared link on any portable system to exchange files with other computers or to access its resources. There are also versions that attach to a printer, letting you send files to the printer using infrared technology.

Complete wireless-LANs are available from companies such as NCR. The NCR WaveLAN PCMCIA network adapter card fits into the mobile user's notebook. It sends and receives RF signals from a network bridge which is connected to the LAN server. PCMCIA stands for the Personal Computer Memory Card International Association. This industry group determines the standards for all credit card-size peripherals, including modems.

A key issue in wireless LANs is security. Most wireless LANs radiate their information in a nonencrypted format, so sensitive information is vulnerable. Encryption should be widely used in the future.

INFRARED CONTROL

Infrared (IR) technology can be embedded into portable devices such as notebook computers, personal digital assistants (PDAs), wireless phones and digital cameras. It can offer valuable HVAC system maintenance features including walk-up interrogation, file transfer and network access.

IR technology in wireless systems offers some advantages over RF technology. These include lower communications subsystem parts cost, smaller physical space and no Federal Communications Commission (FCC) testing and approval requirements. IR-equipped devices also support more secure data transmissions since they provide shorter coverage ranges and IR light cannot escape areas enclosed by walls.

The transceiver is an analog device which converts electrical and optical signals and uses a digital-to-analog converter (DAC), analog-to-digital converter (ADC), preamplifier and gain-control circuits. An object-exchange protocol enables the transfer of files and other objects. There is also serial and parallel port emulation and local area network access for notebook computers and other mobile devices.

The Infrared Data Association (IrDA) Control technology is a command-and-control architecture for communication with wireless peripheral devices. It has an operating range of approximately 7 meters. This system is oriented toward control data packets.

A polled-host topology is used where the host device polls up to eight peripheral devices in an ordered sequence, providing service requests and handling the peripheral-device responses. The host can be personal computers with wireless peripheral devices such as a mouse and keyboard. After the system boots up, the keyboard and mouse operate with the host PC in the same manner as a wired keyboard and mouse.

When an action is to be completed, the controller makes a decision and sends data out through the IR link to the other device. When polled by the host, the device responds, informing the host that it has information to send. The host then requests the information and the device sends it.

The controller passes the data on to the modem function, which handles the coding and modulation. The transceiver performs the electrical-to-optical translation between systems.

PAGERS

Alphanumeric pagers are used for short messages. They usually display a few words or the caller's number. For longer messages, a PCMCIA device can be plugged into a notebook or PDA to receive wireless e-mail.

Wide-area RF equipment may use one-way and two-way data flow. One-way data flow is normally used for digital electronic pagers. In a one-way system, a carrier service, such as SkyTel, transmits a message and an address (the pager ID) over the geographic area it serves. The pager recognizes its ID, then receives and displays the incoming digital message. Two-way service is similar except that the receiving unit also sends data back to the carrier, which forwards the data to another computer on the LAN. There are also PCMCIA cards that can receive and hold pager calls.

Wireless RF modems can be used to connect portable computers to a LAN. A cellular modem or modem/adapter that uses analog audio signals to convey digital information is called a voice-band modem or an analog cellular modem.

Digital cellular modems use digital rather than analog signals to convey information. The advantage is that digital eliminates static, line noise and signal loss during hand-offs. These hand-offs occurs when cellular-based calls are transferred from one cell site to another as the user roams. Roaming occurs when you use a cellular phone in a city other than the one in which you originally set up your account.

RF MODEMS

Radio frequency modems operate in the radio band, which is the portion of the electromagnetic spectrum from 100-kHz to 20-GHz. Wide-area radio-frequency services use an RF modem, which transmits and receives data in packets.

The use of packet data optimizes throughput even under adverse line conditions. The sending modem sends a series of packets, without waiting to see if each is received successfully. If a packet is corrupted by noise or interference, the receiver asks the sending modem to resend the corrupt packet. The sending modem will continue to send new packets until it receives this request. It then resends the errant packet and resumes normal transmission. Since each packet is numbered, they can be resent quickly, even out of order and correctly reassembled by the receiving modem without harming the throughput.

CIRCUIT-SWITCHED CELLULAR

Voice-band cellular or circuit-switched cellular systems use cellular telephone networks, their land lines and microwave technology to make the connection. Cellular phones are part of the mobile radio-telephone system that transmits data over a range of geographic sites, called cells which use low-power transmitters and receivers. The cellular phone can transmit data instead of voice using the audio signals produced by a modem. It is similar to a connection with two land-line-based modems. But, you can remotely access the LAN or call other mobile computers. You substitute a cellular phone for its land-line-based version.

The equipment needed for cellular data transmission is a special type of modem. The cellular modem connects to a cellular phone. Most cellular modems can also use land lines to transmit data and to send and receive fax images.

Cellular channels can have problems with signal quality. Land-based telephone lines offer more consistent quality with minimal noise. When noise is encountered, it is usually due to a faulty connection.

This noise can result in a lower-than-usual transmission rate. Cellular channels are vulnerable to noise and interference which can change in intensity and frequency. Throughput can drop off quickly and the error correction and speed-setting techniques used for land-line telephones will not help.

CELLULAR SECURITY

Cellular telephone traffic, including data from cellular modems, is easy to intercept with inexpensive equipment. In the past even Radio

Shack sold receivers capable of receiving cellular transmissions. Federal laws now prevent the sale of such devices, but many units still remain in use. While the chances of interception may be small, they are still very real. Unless some form of encryption is used, cellular modem users need to be careful about what they send over a cellular phone.

A newer type of wireless technology called CDPD (cellular digital packet data) works by transmitting 128-byte blocks of data over the cellular phone system, looking for unused time in the digital cellular channels network, which is typically idle about 30% of the time. During the idle period, CDPD modems send short bursts of data packets to the cellular network. A cellular adapter connects a standard modem to a cellular phone (via the RJ-11 jack) and enables it to be used for cellular communication.

Unlike voice communications, CDPD automatically includes error-detection and retransmission so that no data are lost. It also has data encryption to secure the transmissions. Another advantage over RF modems and circuit-switched cellular, is that CDPD modems can exchange data with a designated host or service provider or be configured to work with any land-line network and modem. CDPD modems use the same circuitry as digital cellular phones and some can double as digital cellular phones. Some modems are designed to work with portable computers.

WIRELESS LANS

For local area wireless access between two machines, the hardware may be infrared or RF transceivers. Wireless local area networks (WLANs) can use extensions for the 802.11 standard for Ethernet-speed WLANs. Data rates are 20-Mb/s and higher in the 5-GHz frequency range and 10-Mb/s in the 2.4-GHz band.

The modulation scheme used for the 2.4-GHz band is known as complementary code keying (CCK). This technology works in a wide range of different environments, including offices, retail spaces, and warehouses.

At slower speeds, the system shifts into a fallback mode making it possible to increase the coverage areas as it falls back to lower data rates.

This fallback is similar to cell phones, when there is a hand-off from one base station to another as a signal gets weaker and then stronger. In a wireless LAN, you essentially talk to an access point somewhere in the building, connected to the wired network. As you get farther away from it, rather than losing connectivity, it downshifts to a lower data rate, keeping the connection.

WLANs make it possible for multiple users to use the same Internet service provider (ISP) account. The wireless LAN industry is maturing, has standards in place, and is moving ahead with an evolution in products.

POCKET COMMUNICATIONS

Wireless systems devices are evolving to hand-held, multifunctional, portable communicators that combine the capabilities of digital cellular phones with personal digital assistants (PDAs). The latest designs integrate features from multiple communications products including digital cell phones, pagers, data/fax modems, voice messaging and high-end PDAs into a single hand-held unit. Each of these functions is typically handled by separate signal-processing integrated circuits (ICs) called digital-signal-processors (DSPs).

These DSPs provide wired functions such as data/fax modem and wireless functions such as cellular phone, paging, and voice messaging. Telephone-answering with digital recording and playback may also be provided.

A cellular base transceiver station (BTS) serves as the interface between mobile terminals and the rest of the cellular network. The frequency spectrum is allocated in 200-kHz channels, with each channel's bandwidth supporting up to eight users concurrently.

The BTS contains one transceiver for each channel that is allocated to the cell. The BTS must send, receive, and process all of the cellular transmissions in the cell.

The size of a cell depends on the terrain and the amount of calling traffic. Cell sizes range from a maximum radius of approximately 35-km (macrocells) to a few kilometers (microcells) or several hundred meters (picocells). Picocells are needed in densely-populated areas.

In a channel, a time division multiple access (TDMA) format is used so that all users share the same bandwidth. The TDMA has time slots for eight users. The cellular data are transmitted in bursts and each burst fits into an allocated time slot.

The base-station controller (BSC) controls BTS functions for several cells. It hands off communications from one BTS to another as a communicating mobile terminal travels from cell to cell. It also performs speech coding and rate adoption to seamlessly connect the calls between the cellular network and the public-switched (land-based) network.

NETWORK SECURITY

Network security is important in facilities automation. An unauthorized access means more than illegal data manipulation or file corruptions. In real time systems such as building automation and powerplant control systems, an unauthorized command to start or stop expensive equipment many result in severe damage to the equipment and injury to workers.

Control systems should reside on a secured facilities network. The real time network can be connected to the facilities backbone via a network bridge and fire wall to prevent unauthorized access to the nodes on the network.

Facilities need to establish long term support for their systems. A trained and educated staff can assure continuity of systems and network operations. A network manager should be part of the implementation to provide system support and maintain systems security.

Connections to the Internet, dial-up capabilities and the coexistence of database systems for maintenance management systems with real time systems for control and automation on the same network highlight the importance of security.

BAS INTEGRATION AND STANDARDIZATION

In facilities automation, there may be several existing BAS. The integration needs to include:

•   interfaces between systems using open protocols for integration,

- interfaces between BAS and application specific controllers (chillers, dampers, VSD controllers),

- upgrade of systems that are too costly to maintain or integrate; retrofit with new BAS controllers that are networked,

- use standard protocols, implement BACnet or competitive systems.

Standardization of systems components is the key to successful facilities networks. Items for standardization include controls and automation systems, application software, servers, operator workstations, operating systems, communications interfaces and software (Ethernet, ARCNET, RS-485), communication cables, protocols or drivers (BACnet, Profibus, MODBUS) and operator interfaces.

References
Ricketts, Jana, Editor, *Energy and Environmental Visions for the Millennium*, Lilburn, GA: The Fairmont Press, Inc., 1998.

Sioros, Donna, Editor, *Building for the 21st Century: Energy and the Environment*, Lilburn GA: The Fairmont Press, Inc., 2000.

Internet: www.office.com/global "Power Monitoring Software can Help Large Energy Consumers Monitor and Trim Electricity Expenses," July 25, 2001.

Index